The Quantum Nexus: AI, Blockchain, and the Future of Everything

How these Cutting-edge Technologies will Converge to Reshape Various Industries and Everyday Life.

By Dr Israel Carlos Lomovasky

Copyright(c)Dr Israel Carlos Lomovasky 2024

About the Book

Unlocking the Future: Introducing "The Quantum Nexus: AI, Blockchain, and the Future of Everything"

In an era where technology is advancing at an unprecedented pace, understanding the intersection of cutting-edge innovations is crucial for staying ahead. Enter "The Quantum Nexus: AI, Blockchain, and the Future of Everything" – a comprehensive, future-focused exploration of how quantum computing, artificial intelligence, and blockchain technology will converge to reshape industries and everyday life.

A Visionary Guide

"The Quantum Nexus" delves into the transformative potential of these groundbreaking technologies, offering readers a holistic view of the tech landscape from today to the next 10-20 years. This book is designed to be both comprehensive yet accessible, making complex concepts understandable for a broad audience.

Key Features

Comprehensive Coverage: The book covers the fundamentals and advancements of AI, quantum computing, and blockchain technology, providing a solid foundation for readers new to these topics.

Future-Focused Insights: Through case studies, practical examples, and visionary predictions, "The Quantum Nexus" illustrates the potential impacts and uses of these technologies across various sectors, including healthcare, finance, urban planning, and public policy.

Ethical Considerations: The book addresses the ethical and societal implications of technological advancements, offering strategies to ensure that innovation benefits society as a whole.

Practical Guidance: For businesses and individuals, the book provides actionable tips and strategies to prepare for and leverage these emerging technologies, fostering a culture of continuous learning and innovation.

Explore the Intersection of Technologies

Quantum Computing: Learn how quantum computing is set to revolutionize fields like cryptography, drug discovery, and complex system simulations, achieving breakthroughs that were previously unimaginable.

Artificial Intelligence: Discover how AI is transforming industries through advancements in machine learning, neural networks, and natural language processing, driving efficiency and innovation.

Blockchain Technology: Understand the power of blockchain to secure data, enhance transparency, and create decentralized systems, paving the way for a more secure and trustworthy digital world.

Why Read "The Quantum Nexus"?

Stay Informed: Keep up with the latest technological trends and developments, ensuring you are well-informed about the future landscape of these transformative technologies.

Drive Innovation: Gain insights into how to foster a culture of innovation within your organization, encouraging experimentation and collaboration to stay ahead in a competitive market.

Prepare for the Future: Learn practical strategies to future-proof your career and business, from lifelong learning and skills development to strategic planning and ethical technology use.

Engage with Experts: Benefit from the knowledge and perspectives of industry leaders and futurists, who share their visions and predictions for the future.

Join the Technological Revolution

"The Quantum Nexus: AI, Blockchain, and the Future of Everything" is more than just a book – it's a roadmap to the future. Whether you are a business leader, a tech enthusiast, or someone curious about the impact of these technologies, this book offers valuable insights and guidance.

Prepare to unlock the future and navigate the complexities of the technological revolution with confidence. Embrace the convergence of AI, quantum computing, and blockchain, and discover how you can shape the future of innovation.

Get your copy of "The Quantum Nexus" today and embark on a journey into the future of technology!

Stay ahead, drive innovation, and embrace the future with "The Quantum Nexus: AI, Blockchain, and the Future of Everything." Get ready to be inspired and empowered as you explore the transformative potential of these groundbreaking technologies.

Table of Contents

Key Elements:

1. **Introduction to Quantum Computing:**
   - Simplified explanations of quantum mechanics and computing.
   - Current state of quantum computing and major players in the field.

2. **AI Revolution:**
   - Overview of artificial intelligence and machine learning.
   - Current advancements and future potential.

3. **Blockchain and Decentralization:**
   - Basics of blockchain technology and cryptocurrencies.
   - Impacts on finance, supply chain, and data security.

4. **Convergence of Technologies:**

- How quantum computing will enhance AI capabilities.
- Role of blockchain in securing AI-driven systems.

5. **Applications and Industry Impacts:**

    - Case studies on healthcare, finance, urban planning, and public policy.
    - Real-world examples and potential future scenarios.

6. **Ethical and Societal Implications:**

    - Ethical considerations of advanced technologies.
    - Impacts on jobs, privacy, and societal structures.

7. **Future Predictions:**

    - Visionary predictions on the next 10-20 years.
    - Insights from industry leaders and futurists.

8. **Practical Guide for Businesses and Individuals:**

    - How businesses can prepare and leverage these technologies.
    - Tips for individuals to stay ahead in a rapidly evolving tech landscape.

Unique Advantage Points:

- **Comprehensive Yet Accessible:** Balances depth and accessibility, making complex topics understandable to a wide audience.
- **Future-Focused:** Provides a forward-looking perspective that intrigues and inspires readers about the future of technology.

- **Real-World Applications:** Uses case studies and practical examples to illustrate potential impacts and uses.
- **Interdisciplinary Approach:** Integrates multiple cutting-edge technologies, offering a holistic view of the tech landscape.

This book would cater to tech enthusiasts, professionals, policymakers, and anyone interested in the future of technology and its societal impact. By addressing current trends and future possibilities, it has the potential to resonate with a broad audience and become a best-seller in the field of cutting-edge technology.

About the author

Curriculum Vitae

Education:

- Doctor of Science (DSc) in Project Evaluation, Technion, Haifa, Israel
- Master of Science (MSc) in Operations Research, London School of Economics
- Bachelor of Science (BSc) in Industrial and Management Engineering, Technion, Haifa, Israel

Teaching and Academic Research Positions Held:

- Micro Economics
- Macro Economics
- Econometrics
- Statistics
- Mathematics
- Public Finance
- Urban Planning Mathematical Models
- Transportation Science

Urban and Regional Planning Experience:

- Comprehensive Urban Renewal Project Manager (Physical and Social Project) of the East Acco Government Project. Received the title Yakir Acco from the Acco municipality.

Professional Experience:

- Founding partner (2006-2011) in the company "Kaul and Lomovasky Holdings Inc" specializing in the computerization of trading using artificial intelligence.
- Internet and Artificial Intelligence Programmer, Developer, and Consultant (2012-2018).
- Developed an AI-based system to calculate the price of apartments in 300 towns in Israel, using VBA Excel Neural Networks (artificial intelligence) pre-processing and presented the prices on a Python Django-based website.
- (2018-2024)Author of several books on topics such as algorithmic trading, quantum computing, crypto trading, artificial intelligence, and startup ideas.

Computer Programming Skills:

- C, VBA under Excel, Microsoft Office, HTML, PHP, MATLAB, SAS, Python, Django, Keras, Panda, Cloud AI Applications, TensorFlow, Google Cloud Platform, OpenCV, Adversarial GANs, Computer Vision, Image Classification, Object Recognition, Pose Recognition.

- Quantum computing and quantum machine learning.
- Algorithm development, end-to-end ownership.

Publications

-Coding the Citizen's Voice: Python Tools for MOTMSDD in Governance and Planning: the Manual: Python Source Code. AI & Data Science. Metaverse of the Minds ... and Brain Computer Interface Book 9) Kindle Edition
by Dr Israel Carlos Lomovasky (Author) Format: Kindle Edition

Book 8 of 8: The future implications of the combination between the Internet, the Metaverse and Brain Computer Interface

-Beyond the Vote: AI Applications in Direct Democracy and Civic Engagement: Integrating AI, ML, NLP, Data Visualization, and MOTMSDD Into Public Governance ... and Brain Computer Interface Book 8) Kindle Edition
by Dr Israel Carlos Lomovasky (Author) Format: Kindle Edition

Beyond Quantum: The Next Leap in Computational Paradigms: Exploring the Future of Advanced Computing Technologies (Quantum Computing Book 5) Kindle Edition
by Dr Israel Carlos Lomovasky (Author) Format: Kindle Edition

Book 5 of 5: Quantum Computing

-AI-Proof Your Career: Building Resilience in the Face of Automation: Strategies for Healthcare,Finance,Manufacturing,Art,Entertainment,Retail,Transportation,Energy,Logistics,Government,Teaching Kindle Edition
by Dr Israel Carlos Lomovasky (Author) Format: Kindle Edition

-Defensive Trading in Crypto ETFs: Protecting Your Portfolio in Volatile Markets: The Damage and Losses Control Bible for The Crypto ETFs Investor and Trader Kindle Edition
by Dr Israel Carlos Lomovasky (Author) Format: Kindle Edition

Book 11 of 11: TRADING

-Algorithmic Trading for Everyone: A Non-Programmer's Journey to Automation: Comprehensive Introduction to Algo Trading for Beginners Without Programming Background Kindle Edition
by Dr Israel Carlos Lomovasky (Author) Format: Kindle Edition

Book 10 of 10: TRADING

-The Great Crypto Illusion: Navigating the Future of Valueless Assets : Examining the Sustainability of Cryptocurrencies Without Traditional Intrinsic Value. (FINANCE Book 8) Kindle Edition
by Dr Israel Carlos Lomovasky (Author) Format: Kindle Edition

-Navigating Crypto ETFs Trading: An Absolute Beginners Guide to New Markets: Foundations of Crypto ETF Trading: Building Your Digital Investment Portfolio Kindle Edition
by Dr Israel Carlos Lomovasky (Author) Format: Kindle Edition

-Profit and Protect: Retail Trading Strategies to Manage Risk and Grow Your Wealth: Foundations to Advanced. Stocks, Bonds, Crypto, Commodities & Forex. Hedging with Options, Swaps, Futures & More Kindle Edition
by Dr Israel Carlos Lomovasky (Author) Format: Kindle Edition

-The Future Game: Unleashing AI and Quantum Computing Power in Game Theory.: Beginners to Advanced.Python Code.Case studies:Economics,Finance,Politics,Environment,Social Science,Psychology,Health,More Kindle Edition
By Dr Israel Carlos Lomovasky (Author) Format: Kindle Edition

-AI and Quantum Strategies: Python's Role in Economic Innovation: Foundations to Advanced. With python and Quantum Code in a Computational Economics Range of Case Studies Kindle Edition
by Dr Israel Carlos Lomovasky (Author) Format: Kindle Edition

-Quantum Computing in Finance: Bridging Theory and Practice with Python: Case Studies: Algorithmic Trading, Risk Management, Fraud Detection, Options Pricing ,Economic Forecasting and more
by Dr Israel Carlos Lomovasky (Author)

Book 6 of 6: FINANCE

-Artificial Gods: The Onset of Superior Machine Intelligence and Consciousness: : The Why and How of a Ban on Research Leading To Superintelligence And AI Consciousness Kindle Edition
by Dr Israel Carlos Lomovasky (Author)

-Quantum and Consciousness: Exploring the Mind-Computer Interface: Unveiling the Quantum Mind: Quantum Computing and the Fabric of Consciousness Kindle Edition
by Dr Israel Carlos Lomovasky (Author)

-Quantum Democracy: Unleashing MOTMSDD with Quantum Computing: MOTMSDD : Metaverse Of The Minds Social Direct Democracy (The future implications of the ... and Brain Computer Interface Book 6) Kindle Edition
by Dr Israel Carlos Lomovasky (Author)

-MOTMSDD: Metaverse Of The Minds Social Direct Democracy: Governance and Public Decision Making in The Era of Brain Computer Interface, AI and Metaverse, ... and Brain Computer Interface Book 5) Kindle Edition
by Dr Israel Carlos Lomovasky (Author)

-MOTMSDD Urbanism:Redefining Cities through AI and Metaverse of the Minds Social Direct Democracy: Sustainable Urbanism in the Age of Brain-Computer Interface.Solving Conflicts between Citizen's Needs Kindle Edition
by Dr Israel Carlos Lomovasky (Author)

-AI in Financial Markets: A Guide to Algorithmic Trading with ChatGPT: Python Code. CHATGPT Assistance. Basics to Advanced. Traditional and AI/ML Trading. (FINANCE Book 6) Kindle Edition
by Dr Israel Carlos Lomovasky (Author)

-Python for Financial Freedom: Algorithmic Strategies for Personal Wealth: Trading and Investing. Foundations to Advanced. AI/ML, Risk ,Tax ,and Money Management. Stocks & Crypto (FINANCE Book 5) Kindle Edition
by Dr Israel Carlos Lomovasky (Author)

-Quantum Foundations of Computer Vision: A Guide for Researchers and Practitioners: Python and Quantum Language Code. Future Proof Computer Vision (Quantum Computing Book 3) Kindle Edition
by Dr Israel Carlos Lomovasky (Author)

-MOTMSDD ECONOMICS: From Classical Economics, to Metaverse Of The Minds Social Direct Democracy Economics.: For The Next WELFARE ECONOMICS: Harnessing BCI ... the Metaverse . (FUTURE ECONOMICS Book 1) Kindle Edition
by Dr Israel Carlos Lomovasky (Author)

-Quantum Hedge: Unlocking the Future of Algorithmic Trading. : Python and Quantum Languages Code. Basics to Advanced. Stocks, Forex and Crypto. Theory and Hands on Practice. Kindle Edition
by Dr Israel Carlos Lomovasky (Author)

-Quantum Economics: Rethinking Macro and Micro in the Age of Quantum Computing: Theory and Practice: Python and Quantum Language Code Explained Step by Step (FUTURE ECONOMICS Book 2) Kindle Edition
by Dr Israel Carlos Lomovasky (Author)

-Driving with the Mind: Exploring MOTMSDD and Its Impact on Smart Cities and Autonomous Mobility: MOTMSDD: Metaverse of The Minds Social Direct Democracy: ... Meets The Metaverse (URBANISM Book 4) Kindle Edition
by Dr Israel Carlos Lomovasky (Author)

-AI in Fundamental Analysis: Uncovering Hidden Algorithmic Investment Opportunities with Python.: Machine,Reinforcement and Deep Learning.Complete AI-Driven ... Advanced.Risk Management. (FINANCE Book 2) Kindle Edition
by Dr Israel Carlos Lomovasky (Author)

-Python for AI and Creativity: Unleashing the Power of Artificial Intelligence in the Arts: Basics to Advanced.Visual Arts,Design,Music,Poetry,Storytelling, ... learning-Python Book 3) Kindle Edition
by Dr Israel Carlos Lomovasky (Author)

-Python for Machine Learning. From Intermediate to Advanced Guide With Code.: Unleash the Potential of Advanced Machine Learning in Python. Covering Many ... learning-Python Book 2) Kindle Edition
by Dr Israel Carlos Lomovasky (Author)

-Python for Smart Cities: Machine Learning and Artificial Intelligence Applications for Urban Planning and Infrastructure: Python in Action: ML/AI for Smart ... Infrastructure Management (URBANISM Book 2) Kindle Edition
by Dr Israel Carlos Lomovasky (Author)

-Python for Machine Learning: A Beginner's Guide.From Scratch to intermediate.: Basis For Algorithmic Finance, Trading, Healthcare, Industry, Transportation, ... learning-Python Book 1) Kindle Edition
by Dr Israel Carlos Lomovasky (Author)

-SINGULARITY'S VEIL: THE RISE AND FALL OF HUMANITY. : A TALE BETWEEN SCIENCE FICTION AND FUTUROLOGY. STOP ARTIFICIAL GENERAL INTELLIGENCE NOW. (Future sciences - Futurology - Science fiction Book 6) Kindle Edition
by Dr Israel Carlos Lomovasky (Author)

-KILLING THE BEAST. THE THREAT OF ADVANCING ARTIFICIAL GENERAL INTELLIGENCE.: A CALL TO BAN AGI.SURVIVAL OF HUMANITY ON THE LINE. A CONTRARIAN NARRATIVE ... - Futurology - Science fiction Book 5) Kindle Edition
by Dr Israel Carlos Lomovasky (Author)

-Day Trading Basics to Advanced:A Comprehensive Guide.From Scalping to AI/ML.Algorithmic Trading.Python Code.: Day Trading Decoded:Unlocking Secrets to Profitable Trading.Stocks,Crypto,Options,Forex Kindle Edition
by Dr Israel Carlos Lomovasky (Author)

-BEGINNER'S MACHINE LEARNING AND ARTIFICIAL INTELLIGENCE IN PYTHON FOR FINANCE: A GUIDE.: EXPLORING THE INTERSECTION OF FINANCE AND ML/AI: A PYTHON PRIMER Kindle Edition
by Dr Israel Carlos Lomovasky (Author)

-The Internet Of Minds (IOM). An Essay: The Future Implications of Brain Computer Interface
by Dr Israel Carlos Lomovasky (Author)

-CRYPTO TRADING TECHNICAL ANALYSIS: Apply the technical analysis indicators, time-frames and approaches that fit Crypto Currencies trading characteristics. Kindle Edition
by Dr Israel Carlos Lomovasky (Author)

-QUANTUM MACHINE LEARNING: A COMPREHENSIVE GUIDE WITH PRACTICAL EXAMPLES AND QUANTUM LANGUAGE IMPLEMENTATION: FROM BASICS TO ADVANCED.INCLUDES PYTHON CODE. (Quantum Computing Book 2) Kindle Edition
by Dr Israel Carlos Lomovasky (Author)

-CRYPTO BASICS TO ADVANCED. MAKE MONEY WITH CRYPTO.THE CRYPTO BUSINESS STARTUP BIBLE.: Investing ,trading and beyond. 20 Cryptocurrency profitable strategies. Over 100 startup ideas. Kindle Edition
by Dr Israel Carlos Lomovasky (Author)

-QUANTUM COMPUTING AND OPERATIONS RESEARCH.AN ESSAY.WHAT IS QC AND WHY IT MATTERS TO OR PRACTITIONERS.: THE FUTURE IMPLICATIONS OF QUANTUM COMPUTING ON OPTIMIZATION AND OPERATIONS RESEARCH. Kindle Edition
by Dr Israel Carlos Lomovasky (Author)

-ALGORITHMIC TRADING FROM SCRATCH TO AI/ML STRATEGIES IMPLEMENTED IN PYTHON.FOR CRYPTO,STOCKS,FOREX AND MORE.: RETAIL TRADING SYSTEMS FROM BASIC TO SOPHISTICATED STEP BY STEP. PYTHON FOR YOUR PROJECTS. Paperback – May 17, 2023
by Dr Israel Carlos Lomovasky (Author)

-CRYPTO SENTIMENT ALGO TRADING.PYTHON AND PSEUDO-CODE.: Algo Cryptocurrencies Trade: day, trend, news, swing, arbitrage, bots, contrarian, volume, event, seasonal ,and more strategies. Kindle Edition
by Dr Israel Carlos Lomovasky (Author)

-ALGORITHMIC TRADING STRATEGIES AND TECHNIQUES IN PYTHON, PSEUDO-CODE AND TRADESTATION CODE.: Get your projects started.20 most used techniques and strategies covering all tradeable assets. Kindle Edition
by Dr Israel Carlos Lomovasky (Author)
-ALGORITHMIC TRADING STRATEGIES AND TECHNIQUES IN PYTHON, PSEUDO-CODE AND TRADESTATION CODE.: Get your projects started.20 most used techniques and strategies covering all tradeable assets. Kindle Edition
by Dr Israel Carlos Lomovasky (Author)

-

Chapter 1: Introduction to Quantum Computing

1.1 Understanding Quantum Mechanics

1.1.1 The Quantum World

The study of the quantum world represents a significant departure from the principles of classical physics, providing a new lens through which we can understand the fundamental nature of the universe. This section explores the key differences between classical and quantum physics, introduces core quantum concepts, and explains the unique properties of quantum particles.

Classical vs. Quantum Physics

Classical Physics: Classical physics, rooted in the work of Newton, Maxwell, and others, describes the macroscopic world around us. It is deterministic, meaning that the future behavior of a system can be precisely predicted if its initial conditions are known. Classical mechanics deals with continuous variables where quantities such as position, velocity, and energy can take any value within a given range.

Quantum Physics: Quantum physics, on the other hand, governs the behavior of particles at the atomic and subatomic levels. It is inherently probabilistic, dealing with discrete quantities and probabilistic outcomes. Quantum mechanics challenges classical notions with phenomena like superposition, entanglement, and the observer effect, which have no analogs in classical physics.

Key Differences:

- **Determinism vs. Probability:** Classical physics is deterministic, while quantum physics is probabilistic.
- **Scale:** Classical physics applies to macroscopic objects; quantum physics applies to microscopic particles.
- **Continuous vs. Discrete:** Classical physics deals with continuous variables; quantum physics deals with quantized, discrete variables.
- **Measurement:** In classical physics, measurements do not affect the system. In quantum physics, measurement affects the state of the system, known as the observer effect.

Key Concepts in Quantum Mechanics

Superposition: Superposition is a fundamental concept in quantum mechanics where a particle exists simultaneously in multiple states until it is measured. For instance, an electron in a superposition can be in multiple positions at once. When a measurement is made, the electron 'collapses' into one of the possible states.

Entanglement: Entanglement describes a phenomenon where particles become interconnected such that the state of one particle cannot be described independently of the state of the other. Changes to one entangled particle instantly affect the other, regardless of the distance between them. This has profound implications for quantum information and quantum computing.

Quantum Tunneling: Quantum tunneling occurs when particles pass through barriers that would be insurmountable according to classical physics. This is possible because of the wave-like properties of particles in quantum mechanics, allowing them to 'tunnel' through energy barriers.

Quantum Particles: Qubits vs. Classical Bits

Classical Bits: Classical bits are the basic units of information in classical computing, representing either a 0 or a 1. All classical computation relies on these binary states to perform calculations.

Quantum Bits (Qubits): Qubits are the quantum counterparts of classical bits. Unlike classical bits, qubits can exist in a superposition of states, meaning they can represent both 0 and 1 simultaneously. This property, along with entanglement, allows quantum computers to process a vast amount of information in parallel, offering exponential increases in computational power for certain tasks.

Representation and Information Capacity:

- **Classical Bits:** A single classical bit can only be in one state at a time (0 or 1).
- **Qubits:** A single qubit can be in a superposition of both states simultaneously. When multiple qubits are entangled, the information capacity grows exponentially, enabling quantum computers to solve problems that are currently infeasible for classical computers.

By understanding these fundamental concepts, we lay the groundwork for exploring how quantum mechanics not only redefines our understanding of the physical world but also how it intersects with and enhances other cutting-edge technologies like AI and blockchain. The following chapters will delve deeper into these intersections and their implications for various industries and aspects of daily life.

This exploration sets the stage for a future where quantum computing, AI, and blockchain converge to drive unprecedented technological and societal advancements.

1.1.2 The Basics of Quantum Mechanics

Quantum mechanics fundamentally reshapes our understanding of the physical universe, introducing concepts that challenge classical intuitions. This section delves into the core principles of quantum mechanics, including wave-particle duality, the uncertainty principle, and the role of quantum states and wave functions.

Wave-Particle Duality

One of the most revolutionary ideas in quantum mechanics is wave-particle duality, which posits that particles such as electrons and photons exhibit both wave-like and particle-like properties. This duality is illustrated through several key experiments:

The Double-Slit Experiment: In the double-slit experiment, a beam of particles is directed at a barrier with two slits. When only one slit is open, particles behave as expected, forming a pattern directly behind the slit. However, when both slits are open, an interference pattern emerges on the detection screen, resembling the pattern created by waves. This result implies that each particle passes through both slits simultaneously, interfering with itself as a wave would.

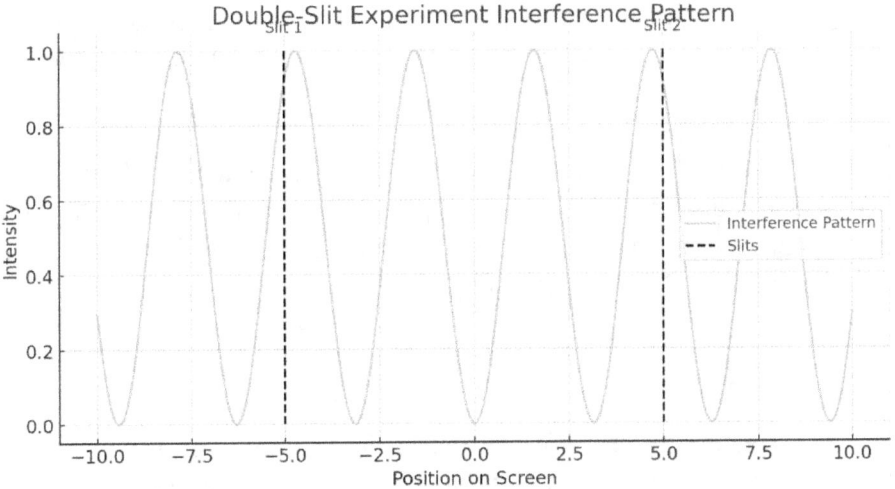

Here is the diagram of the double-slit experiment showing the interference pattern formed by particles passing through two slits. This image illustrates the interference pattern created when particles, such as electrons or photons, pass through two slits and

produce a wave-like interference pattern on the detection screen. This fundamental experiment demonstrates the wave-particle duality in quantum mechanics..

Photon Experiments: Similar experiments with photons demonstrate that light, which was traditionally considered a wave, also exhibits particle-like behavior. When photons are emitted one at a time, they still produce an interference pattern over time, reinforcing the concept of wave-particle duality.

The principle of wave-particle duality reveals that quantum entities cannot be described solely as particles or waves but as a combination of both, depending on the context of the measurement.

The Uncertainty Principle

Heisenberg's uncertainty principle is a cornerstone of quantum mechanics, formulated by Werner Heisenberg in 1927. It states that certain pairs of physical properties, such as position and momentum, cannot be simultaneously measured with arbitrary precision. The more accurately one property is measured, the less accurately the other can be known.

Mathematical Expression: The uncertainty principle is often expressed mathematically as: $\Delta x \cdot \Delta p \geq \frac{h}{4\pi}$ $\Delta x \cdot \Delta p \geq \frac{h}{4\pi}$ where Δx Δx is the uncertainty in position, Δp Δp is the uncertainty in momentum, and h h is Planck's constant.

Implications:

- **Measurement Limitations:** This principle imposes fundamental limits on what can be known about a quantum system. For instance, knowing the exact position of an electron makes its momentum highly uncertain and vice versa.
- **Quantum Systems:** In quantum systems, this principle ensures that particles do not have well-defined paths as in

classical mechanics but rather probabilistic distributions of possible positions and momenta.

Quantum State and Wave Function

The quantum state of a particle encapsulates all its properties and is described by a mathematical function called the wave function, typically denoted by ψ. The wave function provides a probabilistic description of the particle's properties.

Wave Function (ψ): The wave function ψ is a complex-valued function that contains all the information about a quantum system. The probability of finding a particle in a particular state is given by the square of the absolute value of the wave function, $|\psi|^2$.

Schrödinger Equation: The evolution of the wave function over time is governed by the Schrödinger equation, a fundamental equation in quantum mechanics. The time-dependent Schrödinger equation is expressed as: $i\hbar \partial \psi / \partial t = \hat{H} \psi$ where i is the imaginary unit, $\hbar$ is the reduced Planck's constant, $\partial \psi / \partial t$ is the partial derivative of the wave function with respect to time, and $\hat{H}$ is the Hamiltonian operator representing the total energy of the system.

Probability Interpretation:

- **Position Probability Density:** The probability of finding a particle at a specific position x at time t is given by $|\psi(x,t)|^2$.
- **Normalization:** The wave function must be normalized, meaning the total probability of finding the particle in all space is 1: $\int_{-\infty}^{\infty} |\psi(x,t)|^2 \, dx = 1$

Quantum Superposition: Superposition is a principle where a quantum system can exist in multiple states simultaneously. For example, an electron in a quantum superposition of spin-up and spin-down states will not be in a definite spin state until measured.

Image: Visualization of a wave function depicting probability densities for a particle in a quantum state.

These foundational concepts in quantum mechanics—wave-particle duality, the uncertainty principle, and the quantum state described by the wave function—form the bedrock for understanding more advanced topics in quantum computing, artificial intelligence, and blockchain technology. By grasping these basics, we can better appreciate the profound implications of quantum mechanics and its potential to revolutionize technology and industry in the coming decades.

This detailed exploration of quantum mechanics sets the stage for understanding its intersection with AI and blockchain technologies, offering a comprehensive yet accessible introduction to these complex topics.

1.2 Quantum Computing Fundamentals

1.2.1 Qubits and Quantum Gates

The foundation of quantum computing lies in the unique properties of qubits and the operations performed on them by quantum gates. This section delves into what qubits are, how they differ from classical bits, and how quantum gates are used to manipulate qubits, enabling complex computations that surpass the capabilities of classical computers.

What is a Qubit?

A qubit, or quantum bit, is the fundamental unit of quantum information, analogous to a classical bit in classical computing. However, unlike a classical bit, which can be either 0 or 1, a qubit can exist in a superposition of both states simultaneously. This property is what gives quantum computers their extraordinary computational power.

Properties of Qubits:

- **Superposition:** A qubit can be in a state represented as $|\psi\rangle = \alpha|0\rangle + \beta|1\rangle$, where α and β are complex numbers, and $|\alpha|^2 + |\beta|^2 = 1$. This means the qubit is simultaneously in the 0 and 1 states until measured.
- **Entanglement:** Qubits can become entangled, meaning the state of one qubit is directly related to the state of another, no matter the distance between them. This property is crucial for many quantum algorithms.
- **Quantum Interference:** The probabilistic nature of qubits allows for interference effects, which can be harnessed to amplify correct solutions and cancel out incorrect ones in a quantum computation.

Difference from Classical Bits:

- **Classical Bit:** Can be either 0 or 1.
- **Quantum Bit (Qubit):** Can be 0, 1, or any quantum superposition of these states, allowing for parallel processing of information.

Diagram of a Qubit: To visualize a qubit, we often use the Bloch sphere representation, where any point on the sphere represents a possible state of the qubit.

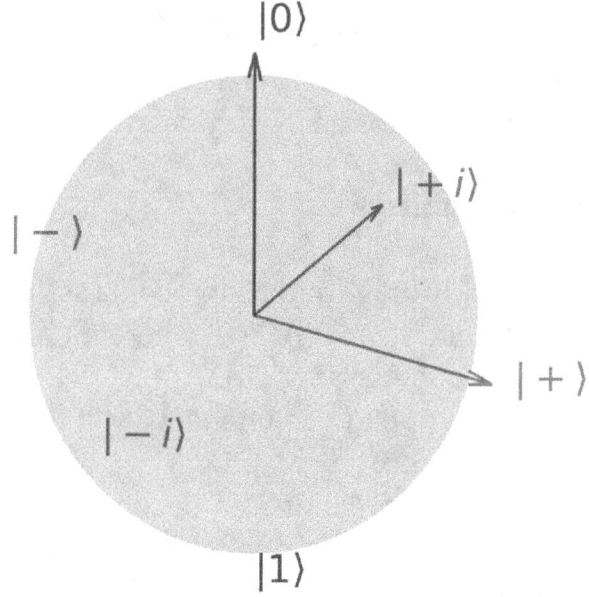

Here is the Bloch Sphere representing a qubit's state. The north and south poles represent the states |0⟩ and |1⟩, respectively, while any point on the surface represents a superposition of these states.

This image helps visualize the complex state space of a qubit, illustrating its capability to exist in multiple states simultaneously, a fundamental property leveraged in quantum computing.

Quantum Gates

Quantum gates are the building blocks of quantum circuits, analogous to logic gates in classical computing. They are operations that change the state of qubits, enabling quantum computations. Unlike classical logic gates, quantum gates are reversible and represented by unitary matrices. Below, we explore some fundamental quantum gates and their operations.

Hadamard Gate (H): The Hadamard gate creates a superposition state from a basis state. When applied to a qubit in the state |0⟩, it transforms the state to 12(|0⟩+|1⟩)21(|0⟩+|1⟩).

$H=12(111−1)H=21(111−1)$

Pauli-X Gate: The Pauli-X gate acts like a classical NOT gate, flipping the state of the qubit. It changes |0⟩ to |1⟩ and vice versa.

$X=(0110)X=(0110)$

Controlled-NOT Gate (CNOT): The CNOT gate is a two-qubit gate where one qubit controls the flipping of the second qubit. If the control qubit is in the state |1⟩, the target qubit is flipped; otherwise, it remains unchanged.

CNOT=(1000010000010010)CNOT= 1000010000010010

Quantum Circuit Example: To illustrate how quantum gates operate in a circuit, consider a simple quantum circuit that creates an entangled state using Hadamard and CNOT gates.

1. **Initial State:** |00⟩
2. **Apply Hadamard Gate to the first qubit:** The state becomes 12(|0⟩+|1⟩)|0⟩21(|0⟩+|1⟩)|0⟩.

3. **Apply CNOT Gate:** The state becomes $\frac{1}{\sqrt{2}}(|00\rangle+|11\rangle)$ ($\frac{1}{\sqrt{2}}(|00\rangle+|11\rangle)$), an entangled state.

Diagram of a Simple Quantum Circuit:

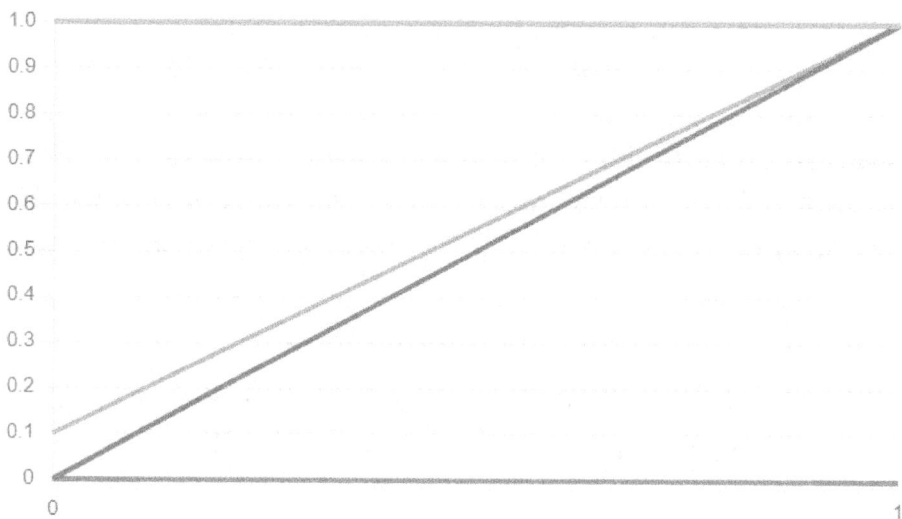

Here is the diagram of a simple quantum circuit showing the initial state |00⟩, the application of a Hadamard gate to the first qubit, and a CNOT gate to create an entangled state.

This image illustrates how quantum gates are used to manipulate qubits and create entangled states, a fundamental operation in many quantum algorithms.

Quantum gates manipulate qubits through these operations, building complex quantum circuits that perform computations far beyond the reach of classical systems. These gates and their interactions form the basis of quantum algorithms, enabling tasks such as factorization, search, and simulation to be performed exponentially faster than with classical computers.

This detailed explanation of qubits and quantum gates introduces the fundamental elements of quantum computing, setting the stage for exploring their applications and implications in subsequent sections and chapters.

1.2.2 Quantum Circuits

Quantum circuits are the fundamental structures in quantum computing, analogous to classical logic circuits but with unique capabilities derived from the principles of quantum mechanics. This subsection delves into the building blocks of quantum circuits and introduces some of the most significant quantum algorithms, highlighting their potential to revolutionize various fields.

Building Blocks

Quantum Gates and Their Combinations: Quantum circuits are constructed using quantum gates, which perform operations on qubits. These gates are the quantum analogs of classical logic gates but operate under the principles of superposition and entanglement. Here are some essential quantum gates and how they are combined to form quantum circuits:

- **Hadamard Gate (H):** Creates a superposition state. When applied to a qubit in the state $|0\rangle$, it produces $\frac{1}{\sqrt{2}}(|0\rangle+|1\rangle)$.
- **Pauli-X Gate (X):** Functions like a classical NOT gate, flipping the state of the qubit from $|0\rangle$ to $|1\rangle$ and vice versa.
- **Pauli-Y and Pauli-Z Gates (Y, Z):** Rotate the state of the qubit around the Y and Z axes of the Bloch sphere, respectively.
- **Controlled-NOT Gate (CNOT):** A two-qubit gate where the state of the second qubit (target) is flipped if the first qubit (control) is in the state $|1\rangle$.
- **Swap Gate:** Exchanges the states of two qubits.

Combining Gates to Form Circuits: To build a quantum circuit, these gates are arranged in sequences to perform specific operations. Each quantum gate corresponds to a unitary matrix, and the sequence of gates can be represented as a product of these matrices. A quantum circuit typically starts with an initial state, applies a series of quantum gates, and ends with a measurement of the qubits.

Example Quantum Circuit: Consider a simple quantum circuit that generates a Bell state (a maximally entangled state) from an initial state $|00\rangle$ using a Hadamard gate and a CNOT gate.

1. **Initial State:** $|00\rangle$
2. **Hadamard Gate on the first qubit:**
   $H|0\rangle = \frac{1}{\sqrt{2}}(|0\rangle+|1\rangle)$ The state becomes $\frac{1}{\sqrt{2}}(|0\rangle+|1\rangle)|0\rangle$.
3. **CNOT Gate with the first qubit as control and the second qubit as target:**
   $CNOT(\frac{1}{\sqrt{2}}(|0\rangle+|1\rangle)|0\rangle) = \frac{1}{\sqrt{2}}(|00\rangle+|11\rangle)$ The resulting state is an entangled Bell state.

Image: A simple quantum circuit showing the initial state $|00\rangle$, the application of a Hadamard gate to the first qubit, and a CNOT gate to create an entangled state.

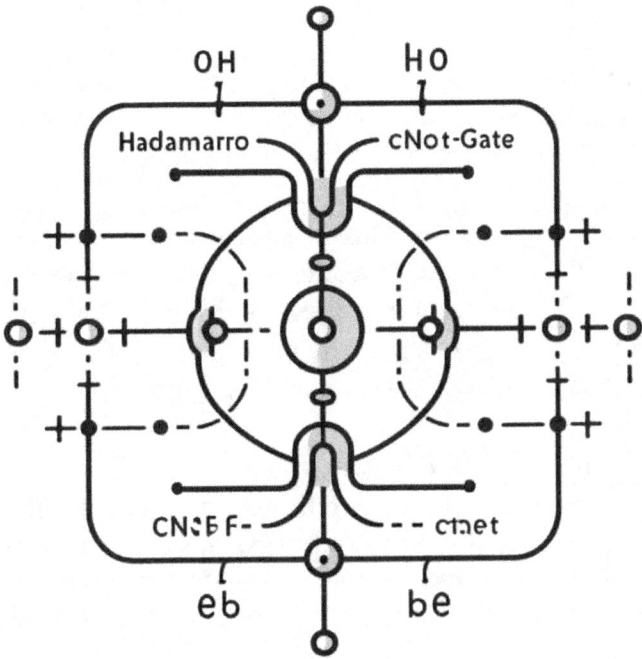

Quantum Algorithms

Quantum algorithms leverage the unique properties of quantum mechanics to solve problems more efficiently than classical algorithms. Here are brief introductions to two of the most famous quantum algorithms:

Shor's Algorithm for Factoring: Shor's algorithm, developed by Peter Shor in 1994, efficiently factors large integers into prime factors. This is significant because the security of many cryptographic systems, such as RSA, relies on the difficulty of factoring large numbers. Shor's algorithm can factor an integer NN in polynomial time, whereas the best-known classical algorithms require exponential time.

- **Quantum Fourier Transform (QFT):** A crucial component of Shor's algorithm, the QFT transforms a quantum state into its frequency domain, enabling the identification of periodicity in the problem.
- **Period Finding:** By exploiting quantum parallelism and interference, Shor's algorithm finds the period of a function, which is used to determine the factors of NN.

Grover's Search Algorithm: Grover's algorithm, developed by Lov Grover in 1996, provides a quadratic speedup for unstructured search problems. If there is an unsorted database of NN items, Grover's algorithm can find a specific item in $O(N)O(N)$ time, compared to $O(N)O(N)$ time for classical algorithms.

- **Amplitude Amplification:** Grover's algorithm uses a process called amplitude amplification to increase the probability amplitude of the desired state, making it more likely to be measured.
- **Oracle Function:** The algorithm queries an oracle function that identifies the desired item and applies a sequence of quantum operations to amplify the correct solution.

Example of Grover's Algorithm:

1. **Initialize the System:** Start with an equal superposition of all possible states.
2. **Apply the Oracle:** Invert the amplitude of the target state.
3. **Amplitude Amplification:** Use the Grover iteration to amplify the amplitude of the target state.
4. **Measurement:** Measure the state, which yields the target item with high probability.

Visualizing Grover's Algorithm:

Image: Visualization of Grover's algorithm showing the process of amplitude amplification.

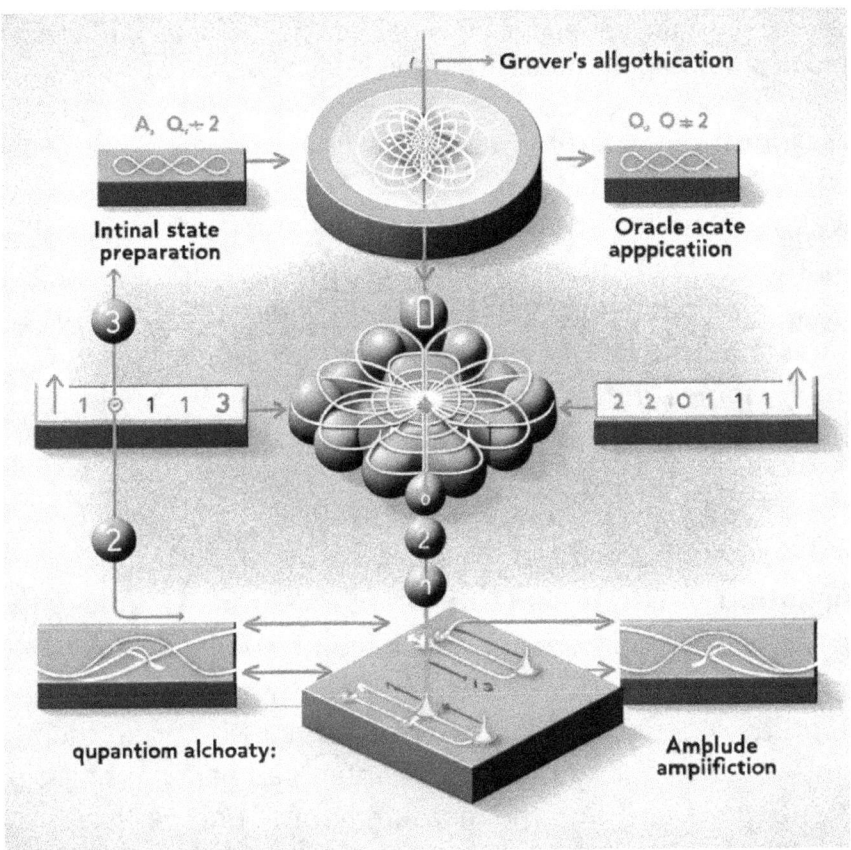

These algorithms demonstrate the power of quantum computing to tackle problems that are currently intractable for classical computers. By understanding the building blocks of quantum circuits and the principles behind these quantum algorithms, we can appreciate the profound potential of quantum computing to revolutionize industries ranging from cryptography to search optimization.

This detailed exploration of quantum circuits and algorithms provides a foundational understanding of how quantum computing operates and its transformative potential, setting the stage for further exploration of its applications in various domains.

1.2.3 Quantum Entanglement

Quantum entanglement is a fundamental and fascinating phenomenon in quantum mechanics, often described as "spooky action at a distance" by Albert Einstein. This section explores the concept of entanglement, its significance in quantum computing, and its practical applications in fields such as quantum cryptography and communication.

Spooky Action at a Distance

Explanation of Entanglement: Quantum entanglement occurs when two or more particles become so deeply linked that the state of one particle instantaneously influences the state of the other(s), regardless of the distance between them. This non-local property means that entangled particles share a joint quantum state, and the measurement of one particle's state will immediately determine the state of its entangled partner(s).

Einstein-Podolsky-Rosen (EPR) Paradox: In 1935, Albert Einstein, Boris Podolsky, and Nathan Rosen proposed a thought experiment, known as the EPR paradox, to challenge the completeness of quantum mechanics. They argued that if quantum mechanics were complete, it would imply that particles could instantaneously affect each other across any distance, which they believed was impossible.

Bell's Theorem and Experiments: John Bell formulated Bell's Theorem in 1964, providing a way to test the predictions of quantum mechanics against those of local hidden variable theories. Experiments conducted since then have overwhelmingly supported the predictions of quantum mechanics, confirming that entanglement is real and that "spooky action at a distance" does occur.

Diagram of Quantum Entanglement: To illustrate quantum entanglement, consider two entangled photons. When Photon A is measured and found to be in state $|0\rangle$, Photon B will instantaneously be found in state $|1\rangle$, and vice versa.

Image: Diagram showing two entangled particles. When the state of one particle is measured, the state of the other is instantaneously determined, regardless of the distance between them.

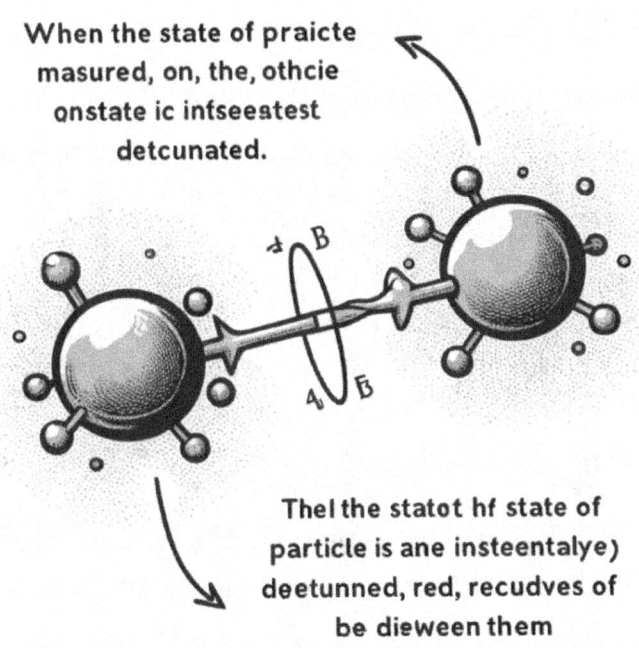

Applications of Entanglement

Quantum Cryptography: Quantum entanglement forms the basis of quantum cryptography, specifically Quantum Key Distribution (QKD) protocols like BB84 and E91. These protocols leverage the principles of entanglement to enable secure communication between parties.

Quantum Key Distribution (QKD): QKD allows two parties to generate a shared, secret cryptographic key by transmitting entangled particles. Any attempt by an eavesdropper to intercept the key will disturb the entangled state, alerting the communicating parties to the presence of the eavesdropper and ensuring the security of the key exchange.

Example of BB84 Protocol:

1. **Preparation:** Alice prepares a series of qubits in randomly chosen basis states (rectilinear or diagonal).
2. **Transmission:** Alice sends these qubits to Bob.
3. **Measurement:** Bob measures each qubit in a randomly chosen basis.
4. **Key Sifting:** Alice and Bob publicly compare their basis choices and discard any bits where their choices do not match.
5. **Error Checking:** They perform error checking to detect any eavesdropping.
6. **Key Generation:** The remaining bits form the secure cryptographic key.

Quantum Communication: Entanglement enables quantum communication protocols that can achieve tasks impossible with classical communication alone. This includes quantum teleportation and superdense coding.

Quantum Teleportation: Quantum teleportation is a protocol that allows the transfer of quantum information from one location to another, using entanglement and classical communication. It does not involve the physical transfer of particles but the transfer of the quantum state.

Steps in Quantum Teleportation:

1. **Entanglement Creation:** Alice and Bob share an entangled pair of qubits.
2. **State Preparation:** Alice has a qubit in an unknown state that she wishes to teleport.
3. **Bell Measurement:** Alice performs a joint measurement on her qubit and her half of the entangled pair, projecting them into a Bell state.
4. **Classical Communication:** Alice sends the result of her measurement to Bob via a classical communication channel.
5. **State Reconstruction:** Bob uses the information received from Alice to apply the appropriate quantum gate to his half of the entangled pair, transforming it into the state of Alice's original qubit.

Diagram of Quantum Teleportation:

Image: Diagram illustrating the steps of quantum teleportation: entanglement creation, state preparation, Bell measurement, classical communication, and state reconstruction.

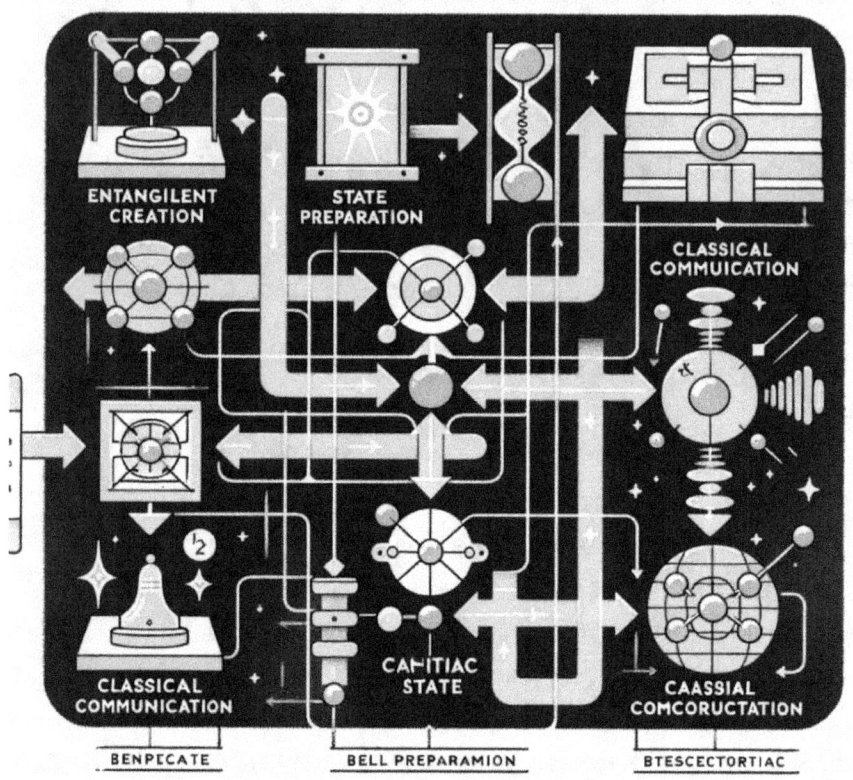

Superdense Coding: Superdense coding allows the transmission of two classical bits of information using only one qubit, leveraging entanglement to increase the information capacity of quantum communication channels.

Steps in Superdense Coding:

1. **Entanglement Creation:** Alice and Bob share an entangled pair of qubits.
2. **Encoding:** Alice encodes two classical bits of information by applying one of four quantum gates (I, X, Y, Z) to her qubit.
3. **Transmission:** Alice sends her qubit to Bob.

4. **Decoding:** Bob performs a Bell state measurement on the received qubit and his half of the entangled pair, determining the two classical bits of information.

Diagram of Superdense Coding:

Image: Diagram showing the process of superdense coding, including entanglement creation, encoding, transmission, and decoding.

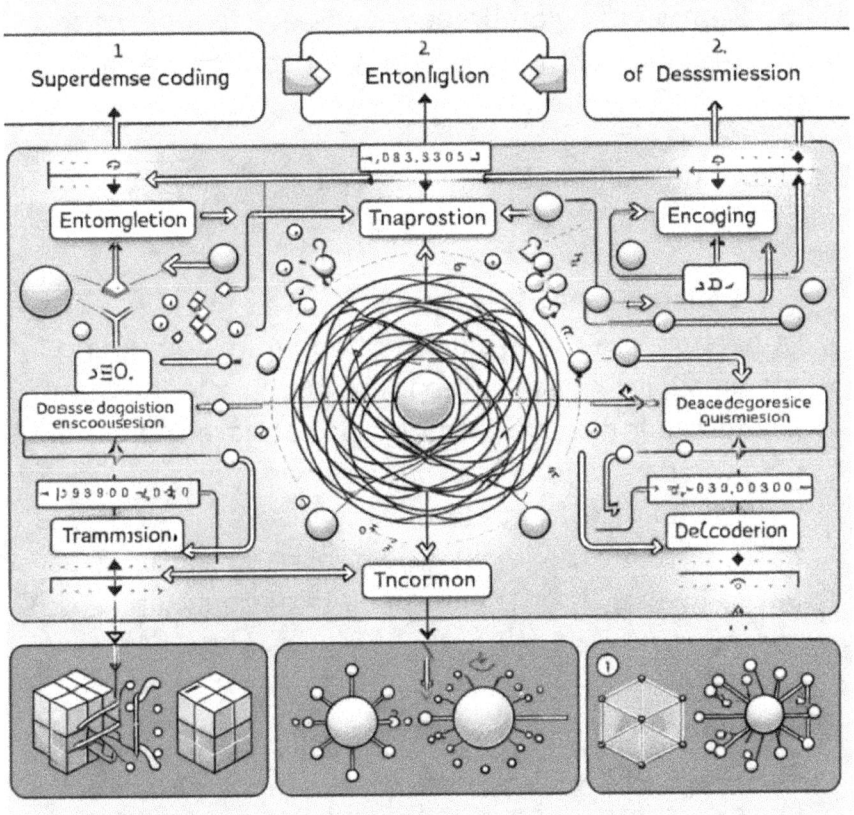

Through these applications, quantum entanglement demonstrates its potential to revolutionize communication and cryptography, offering unparalleled security and efficiency. Understanding entanglement is key to harnessing the power of quantum technologies and realizing their full potential in the next decade and beyond.

This detailed explanation of quantum entanglement and its applications provides a comprehensive understanding of this critical concept in quantum mechanics, illustrating its profound implications for future technologies.

1.3 The State of Quantum Computing Today

1.3.1 Historical Milestones

Understanding the evolution of quantum computing requires a look back at the foundational theories and key technological breakthroughs that have shaped this revolutionary field. This section provides an overview of the pivotal milestones, from early theoretical concepts to significant experimental demonstrations, that have paved the way for the current state of quantum computing.

Early Theories

Max Planck and Quantum Hypothesis (1900): The origins of quantum theory can be traced back to Max Planck's introduction of the quantum hypothesis in 1900. Planck proposed that energy is quantized and can be emitted or absorbed in discrete units called "quanta." This concept was critical in explaining blackbody radiation and laid the groundwork for quantum mechanics.

Albert Einstein and the Photoelectric Effect (1905): Einstein's explanation of the photoelectric effect provided further evidence of quantized energy levels. He proposed that light consists of particles called photons, each carrying a quantum of energy proportional to its frequency. This work earned him the Nobel Prize in Physics in 1921 and reinforced the idea of quantization.

Niels Bohr and the Bohr Model (1913): Niels Bohr developed a model of the atom that incorporated quantized electron orbits. According to Bohr's model, electrons orbit the nucleus in discrete energy levels, and the emission or absorption of light occurs when electrons transition between these levels. This model successfully explained the spectral lines of hydrogen.

Erwin Schrödinger and the Schrödinger Equation (1926): Erwin Schrödinger formulated the Schrödinger equation, a fundamental equation of quantum mechanics that describes how the quantum state of a physical system changes over time. The equation's solutions, known as wave functions, provide a probabilistic description of a particle's position and momentum.

Werner Heisenberg and the Uncertainty Principle (1927): Heisenberg introduced the uncertainty principle, which states that certain pairs of physical properties, such as position and momentum, cannot be simultaneously measured with arbitrary precision. This principle highlights the inherent limitations in our ability to measure and predict quantum phenomena.

Paul Dirac and Quantum Field Theory (1928): Paul Dirac extended quantum mechanics to incorporate special relativity, resulting in the Dirac equation. His work laid the foundation for quantum field theory and predicted the existence of antimatter, leading to significant advancements in theoretical physics.

John Bell and Bell's Theorem (1964): John Bell formulated Bell's Theorem, which provided a way to test the predictions of quantum mechanics against those of local hidden variable theories. Bell's Theorem and subsequent experiments demonstrated the reality of quantum entanglement, a phenomenon that defies classical intuition.

Diagram of Early Quantum Theories: To visually represent these early milestones, consider a timeline highlighting the key theories and discoveries from Planck's quantum hypothesis to Bell's Theorem.

Image: Timeline illustrating the development of early quantum theories from 1900 to 1964.

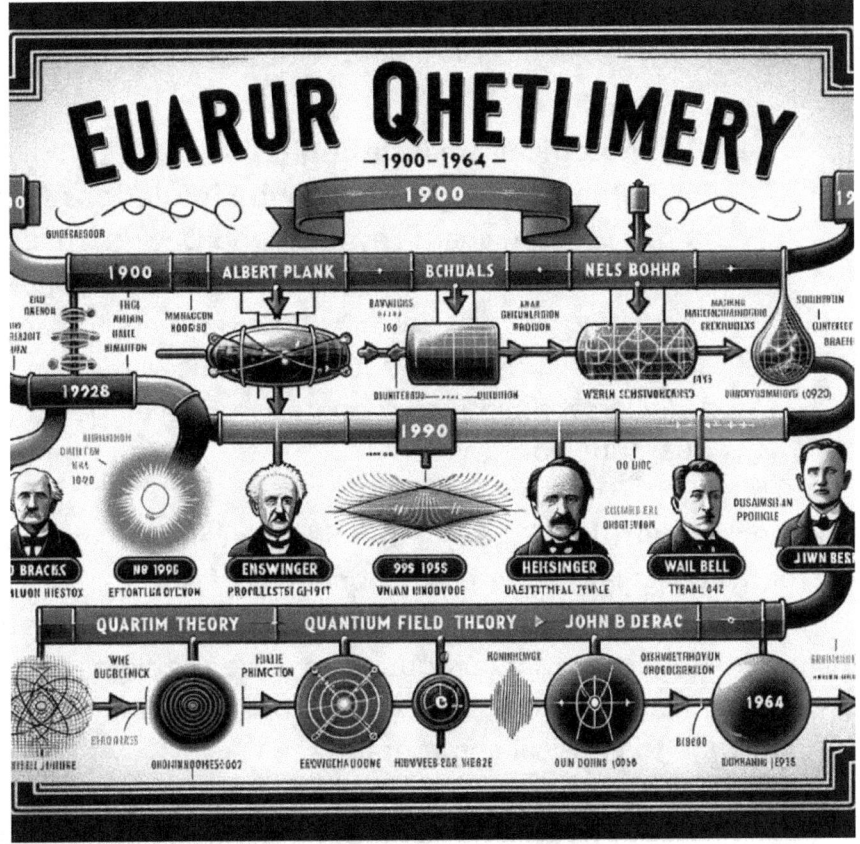

Here is the timeline illustrating the development of early quantum theories from 1900 to 1964.

This image includes key milestones such as Max Planck's quantum hypothesis (1900), Albert Einstein's photoelectric effect (1905), Niels Bohr's atomic model (1913), Erwin Schrödinger's wave equation (1926), Werner Heisenberg's uncertainty principle (1927), Paul Dirac's quantum field theory (1928), and John Bell's theorem (1964).

Technological Breakthroughs
First Quantum Algorithms and Simulations (1980s): In the 1980s, Richard Feynman and David Deutsch proposed the idea of quantum computers capable of simulating physical systems more efficiently than classical computers. Deutsch introduced the concept of a universal quantum computer and developed the first quantum algorithms.

Shor's Algorithm (1994): Peter Shor developed an algorithm for factoring large integers exponentially faster than the best-known classical algorithms. Shor's algorithm demonstrated the potential of quantum computing to solve problems that are intractable for classical computers, posing a significant challenge to classical cryptography.

Grover's Algorithm (1996): Lov Grover introduced an algorithm that provides a quadratic speedup for unstructured search problems. Grover's algorithm further showcased the power of quantum computing to enhance computational efficiency.

Experimental Demonstrations of Quantum Gates (1990s - 2000s): Experimental advancements in the 1990s and 2000s included the successful implementation of basic quantum gates and simple quantum circuits. These experiments demonstrated the feasibility of manipulating qubits and performing quantum operations.

IBM's Quantum Computer and Quantum Volume (2017): In 2017, IBM announced the development of a

20-qubit quantum computer and introduced the concept of quantum volume as a metric for evaluating the performance of quantum computers. Quantum volume considers factors such as the number of qubits, gate fidelity, and error rates.

Google's Quantum Supremacy Claim (2019): In 2019, Google claimed to have achieved quantum supremacy with its 53-qubit Sycamore processor. The quantum computer performed a specific computational task in 200 seconds, which would take the world's fastest supercomputer approximately 10,000 years to complete.

Development of Error-Correcting Codes and Fault-Tolerant Quantum Computing (2020s): Recent advancements have focused on developing quantum error-correcting codes and fault-tolerant quantum computing architectures. These technologies are essential for building scalable and reliable quantum computers capable of solving practical problems.

Diagram of Technological Breakthroughs: A diagram illustrating the timeline of significant technological breakthroughs in quantum computing can provide a visual summary of these milestones.

Image: Timeline highlighting significant technological advancements in quantum computing from the 1980s to the 2020s.

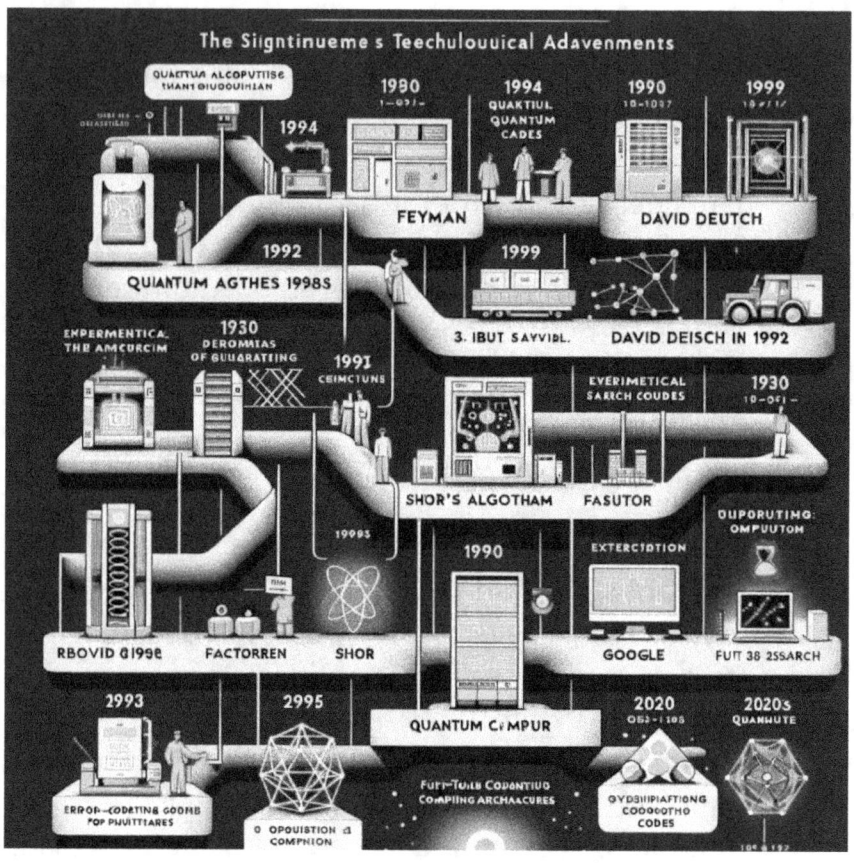

Here is the timeline highlighting significant technological advancements in quantum computing from the 1980s to the 2020s.

This image includes milestones such as the proposal of quantum algorithms by Richard Feynman and David Deutsch in the 1980s, Shor's algorithm for factoring (1994), Grover's search algorithm (1996), experimental demonstrations of quantum gates (1990s-2000s), IBM's 20-qubit quantum computer and the introduction of quantum volume (2017), Google's claim of quantum supremacy (2019), and the development of error-correcting codes and fault-tolerant quantum computing architectures (2020s).

> This detailed overview of historical milestones in quantum computing provides a comprehensive understanding of the theoretical foundations and technological advancements that have shaped this field. By examining these key developments, we can appreciate the progress made and the potential future impact of quantum computing on various industries and everyday life.

1.3.2 Current Quantum Technologies

The landscape of quantum computing is rapidly evolving, driven by advances in both quantum hardware and software. This section explores the various types of quantum computers, detailing the underlying technologies, and provides an overview of the quantum programming languages and frameworks that enable researchers and developers to harness the power of quantum computing.

Quantum Hardware

Superconducting Qubits: Superconducting qubits are one of the most mature and widely used types of qubits in current quantum computers. These qubits are formed using superconducting circuits that can maintain quantum states for significant periods.

- **Principle:** Superconducting qubits operate at temperatures near absolute zero, using Josephson junctions to create and manipulate qubits. These qubits are controlled using microwave pulses.
- **Advantages:** High coherence times, scalability, and compatibility with existing semiconductor fabrication technologies.
- **Challenges:** Requires extremely low temperatures and sophisticated cryogenic systems to maintain superconductivity.

Key Players:

- **IBM:** IBM's quantum computers, such as the IBM Quantum Experience, utilize superconducting qubits. IBM has also introduced the concept of quantum volume as a comprehensive metric for quantum computer performance.
- **Google:** Google's Sycamore processor, which achieved quantum supremacy in 2019, is based on superconducting qubits.

Trapped Ions: Trapped ion quantum computers use ions (charged atoms) confined in electromagnetic traps as qubits. Laser beams are used to manipulate the quantum states of these ions.

- **Principle:** Ions are trapped using electromagnetic fields in a vacuum chamber, and lasers are used to cool the ions and perform quantum operations.
- **Advantages:** High-fidelity quantum gates and long coherence times due to the isolation of ions from the environment.

- **Challenges:** Scalability is limited by the complexity of trapping and controlling large numbers of ions.

Key Players:

- **IonQ:** IonQ has developed quantum computers based on trapped ion technology, known for their high gate fidelity and modularity.
- **Honeywell:** Honeywell's quantum computer uses trapped ions and boasts high quantum volume metrics.

Topological Qubits: Topological qubits aim to encode quantum information in the global properties of particles, such as anyons, which are less susceptible to local noise and errors.

- **Principle:** Topological qubits rely on the braiding of anyons in two-dimensional systems to perform quantum computations. These qubits are theoretically more robust against errors.
- **Advantages:** Potential for intrinsically fault-tolerant quantum computing, which could reduce the overhead needed for error correction.
- **Challenges:** Experimental realization of topological qubits is still in the early stages, with significant theoretical and practical hurdles to overcome.

Key Players:

- **Microsoft:** Microsoft's Quantum program is heavily invested in developing topological qubits through its research on topological insulators and Majorana fermions.

Quantum Software

The development of quantum algorithms and applications requires robust software tools and frameworks. Several quantum programming languages and frameworks have emerged to facilitate this development.

Qiskit: Qiskit is an open-source quantum computing software development framework developed by IBM. It provides tools for creating, simulating, and running quantum circuits on IBM's quantum hardware.

- **Components:**
    - **Qiskit Terra:** The foundation of Qiskit, enabling users to design and optimize quantum circuits.
    - **Qiskit Aer:** Provides high-performance simulators for testing quantum circuits.
    - **Qiskit Ignis:** Focuses on error correction and noise characterization.
    - **Qiskit Aqua:** Facilitates the development of quantum applications in fields such as chemistry, optimization, and machine learning.

Example of Qiskit Code:

```
from qiskit import QuantumCircuit, transpile, Aer, execute



# Create a Quantum Circuit acting on a quantum register of two qubits
```

```python
qc = QuantumCircuit(2)

# Add a H gate on qubit 0
qc.h(0)

# Add a CX (CNOT) gate on control qubit 0 and target qubit 1
qc.cx(0, 1)

# Simulate the quantum circuit
simulator = Aer.get_backend('statevector_simulator')
result = execute(qc, simulator).result()
statevector = result.get_statevector()

print(statevector)
```

Cirq: Cirq is an open-source framework developed by Google for creating, simulating, and executing quantum circuits on Google's quantum processors.

- **Features:**
    - **Quantum Circuits:** Tools for building and manipulating quantum circuits.
    - **Simulation:** High-performance simulators for testing and debugging quantum circuits.
    - **Integration:** Seamless integration with Google's quantum hardware and cloud services.

Example of Cirq Code:

```
import cirq

# Create a quantum circuit with two qubits
qubits = [cirq.GridQubit(0, i) for i in range(2)]
circuit = cirq.Circuit()

# Apply a Hadamard gate on the first qubit
circuit.append(cirq.H(qubits[0]))

# Apply a CNOT gate between the first and second qubit
circuit.append(cirq.CNOT(qubits[0], qubits[1]))

# Simulate the circuit
simulator = cirq.Simulator()
result = simulator.simulate(circuit)

print(result)
```

Other Frameworks:

- **PyQuil:** Developed by Rigetti Computing, PyQuil is a Python library for quantum programming using Quil (Quantum Instruction Language).
- **Microsoft QDK:** The Quantum Development Kit (QDK) by Microsoft includes Q#, a quantum programming language, along with tools and libraries for developing quantum applications.

Diagram of Quantum Software Frameworks:

This detailed exploration of current quantum technologies provides a comprehensive understanding of the different types of quantum hardware and the software tools available for developing quantum applications. By examining these technologies, we can appreciate the advancements made in the field and the potential for future innovations that will shape industries and everyday life.

1.3.3 Major Players and Institutions

The rapid advancements in quantum computing are driven by the concerted efforts of leading companies, academic and research institutions, and government and international initiatives. This section profiles the key players and their contributions to the field, providing a comprehensive overview of the landscape.

Leading Companies

IBM: IBM is a pioneer in quantum computing, making significant contributions through its IBM Quantum Experience platform. IBM's quantum computers use superconducting qubits, and the company has introduced the concept of quantum volume to measure the performance of quantum systems.

- **Key Contributions:**
    - **IBM Q System One:** The world's first integrated universal quantum computing system designed for scientific and commercial use.
    - **Qiskit:** An open-source quantum computing framework that provides tools for creating and executing quantum circuits.

Google: Google has made headlines with its advancements in quantum computing, particularly with its claim of achieving quantum supremacy in 2019. Google's quantum processor, Sycamore, demonstrated the capability to solve a specific problem much faster than the most powerful classical supercomputers.

- **Key Contributions:**
    - **Sycamore Processor:** A 53-qubit quantum processor used to demonstrate quantum supremacy.
    - **Cirq:** An open-source framework for developing, simulating, and executing quantum algorithms.

Microsoft: Microsoft is heavily invested in developing scalable quantum computing solutions through its Azure Quantum platform. The company focuses on a holistic approach that includes hardware, software, and algorithms.

- **Key Contributions:**
    - **Topological Qubits:** Research into topological qubits for robust and fault-tolerant quantum computing.
    - **QDK (Quantum Development Kit):** Includes the Q# programming language and tools for quantum algorithm development.

Rigetti Computing: Rigetti Computing is a startup focused on building full-stack quantum computing solutions. The company develops both quantum hardware and the software needed to operate it.

- **Key Contributions:**
    - **Quantum Cloud Services:** Provides cloud-based access to quantum computers.
    - **PyQuil:** A Python library for quantum programming using the Quil language.

D-Wave Systems: D-Wave Systems specializes in quantum annealing, a different approach from the gate-based quantum computing used by other companies. D-Wave's systems are designed to solve optimization problems.

- **Key Contributions:**
    - **Quantum Annealers:** D-Wave's quantum annealing systems, such as the Advantage

system with over 5,000 qubits, are used for optimization problems.
- **Leap:** A cloud-based quantum application environment that provides access to D-Wave's quantum computers.

Academic and Research Institutions

MIT (Massachusetts Institute of Technology): MIT is at the forefront of quantum research, with significant contributions in quantum algorithms, quantum information theory, and quantum hardware development.

- **Key Contributions:**
    - **Center for Quantum Engineering:** Focuses on developing scalable quantum systems.
    - **MIT IQ (Initiative for Quantum Computing):** Collaborative research program exploring various aspects of quantum computing.

Caltech (California Institute of Technology): Caltech is renowned for its theoretical and experimental research in quantum computing, particularly in quantum error correction and quantum information theory.

- **Key Contributions:**
    - **Institute for Quantum Information and Matter (IQIM):** Conducts research on quantum computation, quantum many-body physics, and quantum optics.

University of Oxford: The University of Oxford has a strong quantum computing program, contributing to quantum algorithms, quantum cryptography, and the development of quantum hardware.

- **Key Contributions:**
    - **Oxford Quantum:** A research group working on various aspects of quantum information and computation.

University of Waterloo: The University of Waterloo, through its Institute for Quantum Computing (IQC), is a global leader in quantum research and education.

- **Key Contributions:**
    - **Quantum Cryptography:** Pioneering research in quantum key distribution and secure communication.
    - **Quantum Algorithms:** Development of new quantum algorithms for various applications.

Government and International Efforts

United States: The U.S. government has launched several initiatives to support quantum research and development, including the National Quantum Initiative Act.

- **Key Programs:**
    - **National Quantum Initiative (NQI):** Aims to accelerate quantum research and development in the U.S.

- **Quantum Economic Development Consortium (QED-C):** Facilitates collaboration between industry, academia, and government.

European Union: The European Union is investing heavily in quantum research through its Quantum Flagship program, a 10-year initiative to support quantum technologies.

- **Key Programs:**
  - **Quantum Flagship:** Focuses on advancing quantum communication, computing, simulation, and metrology.
  - **Horizon Europe:** Includes funding for quantum research projects.

China: China is rapidly advancing its quantum capabilities, with significant investments in quantum communication and computing.

- **Key Programs:**
  - **Quantum Experiments at Space Scale (QUESS):** A satellite-based quantum communication experiment.
  - **National Laboratory for Quantum Information Sciences:** Focuses on developing quantum technologies.

Japan: Japan is investing in quantum research through its Moonshot Research and Development Program, aiming to achieve practical quantum computing by the 2030s.

- **Key Programs:**
    - **Moonshot R&D Program:** Supports ambitious research projects, including quantum computing and quantum communication.

This detailed exploration of the major players and institutions in quantum computing provides a comprehensive overview of the key contributors driving advancements in the field. By examining the efforts of leading companies, academic institutions, and governmental initiatives, we can appreciate the collaborative efforts shaping the future of quantum technologies and their potential impact on various industries and everyday life.

1.4 Challenges and Opportunities in Quantum Computing

1.4.1 Technical Challenges

Quantum computing, despite its immense potential, faces several significant technical challenges that must be overcome to realize practical, large-scale quantum systems. This section delves into three critical challenges: error rates and decoherence, scalability issues, and cryogenic requirements.

Error Rates and Decoherence

Quantum Decoherence: Quantum decoherence is the process by which a quantum system loses its quantum properties due to interaction with its environment. This interaction causes the system to transition from a coherent superposition state to a mixed state, effectively destroying the quantum information.

- **Causes of Decoherence:**
    - **Environmental Interference:** Interaction with external electromagnetic fields, thermal vibrations, and other particles can lead to decoherence.
    - **Intrinsic Noise:** Imperfections in the quantum system itself, such as fluctuations in the control parameters, contribute to decoherence.

Impact on Quantum Computing: Decoherence poses a significant challenge for quantum computing as it limits the time available for performing quantum operations before the quantum information is lost. High error rates due to decoherence reduce the fidelity of quantum computations and make reliable quantum computation difficult.

Error Rates: Quantum gates and operations are susceptible to errors due to various factors, including control inaccuracies and environmental disturbances. The error rate is a measure of the likelihood that a quantum gate operation will fail to produce the correct output.

- **Types of Errors:**
    - **Bit-flip Errors:** A qubit erroneously flips from $|0\rangle$ to $|1\rangle$ or vice versa.
    - **Phase-flip Errors:** A qubit's phase is altered, affecting the superposition state without changing the probabilities of measurement outcomes.
    - **Depolarizing Errors:** A combination of bit-flip and phase-flip errors.

Error Correction: To address the issue of high error rates, quantum error correction techniques are employed. These techniques involve encoding quantum information redundantly using multiple physical qubits to form logical qubits that can detect and correct errors.

Example of Quantum Error Correction: The surface code is a prominent quantum error-correcting code that uses a two-dimensional array of physical qubits to protect quantum information.

Scalability Issues

Scaling Up Quantum Systems: Building a practical quantum computer requires scaling up the number of qubits from the current tens or hundreds to thousands or millions. Several challenges must be addressed to achieve this scalability.

- **Control and Readout:** Controlling and reading out the state of a large number of qubits with high precision is technically demanding.
- **Connectivity:** Ensuring adequate connectivity between qubits to implement complex quantum algorithms requires sophisticated architectures.
- **Error Management:** As the number of qubits increases, so does the complexity of error correction, necessitating efficient and scalable error-correcting codes.

Physical Implementations: Different quantum computing technologies face unique scalability challenges based on their physical implementations.

- **Superconducting Qubits:** Require intricate wiring and control systems to manage large arrays of qubits, posing challenges in minimizing crosstalk and interference.
- **Trapped Ions:** Scaling trapped ion systems involves managing and controlling larger ion traps, which becomes increasingly complex.
- **Topological Qubits:** While potentially offering more straightforward error correction, topological qubits are still in the experimental phase and face challenges in practical implementation.

Cryogenic Requirements

Extremely Low Temperatures: Many quantum computing technologies, particularly those based on superconducting qubits, require operation at cryogenic temperatures close to absolute zero (near 0 Kelvin or -273.15°C). These temperatures are necessary to maintain the superconducting state and reduce thermal noise, which can cause decoherence.

- **Cryogenic Systems:**
    - **Dilution Refrigerators:** Commonly used to achieve the necessary low temperatures, dilution refrigerators use a mixture of helium-3 and helium-4 to reach temperatures in the millikelvin range.
    - **Challenges:** Maintaining such low temperatures requires complex and expensive infrastructure, including cryogenic cooling systems, vacuum chambers, and precise thermal management.

Impact on Practical Deployment: The need for cryogenic environments limits the practicality and portability of quantum computers. Developing systems that can operate at higher temperatures or finding alternative qubit technologies that do not require extreme cooling is an area of active research.

Alternative Approaches: Research is ongoing to develop qubit technologies that can operate at higher temperatures, such as silicon-based qubits and diamond nitrogen-vacancy (NV) centers.

This detailed exploration of the technical challenges in quantum computing provides a comprehensive understanding of the hurdles that must be overcome to realize practical and scalable quantum systems. By addressing issues related to error rates and decoherence, scalability, and cryogenic requirements, researchers and engineers are working towards unlocking the full potential of quantum computing for various applications in the next decade and beyond.

1.4.2 Opportunities for Innovation

Quantum computing holds the potential to revolutionize various industries by providing unprecedented computational power and enabling solutions to problems that are currently intractable for classical computers. This section explores the transformative potential of quantum computing across different fields, as well as the promise of hybrid quantum-classical approaches that leverage the strengths of both paradigms.

Revolutionizing Industries

Cryptography: Quantum computing poses both challenges and opportunities for cryptography. While it threatens to break widely used cryptographic systems, it also opens the door to new, quantum-resistant cryptographic methods.

- **Breaking Traditional Cryptography:** Shor's algorithm can factor large integers exponentially faster than classical algorithms, potentially breaking RSA and other encryption schemes.
- **Quantum Key Distribution (QKD):** QKD, based on the principles of quantum mechanics, enables secure communication channels that are theoretically immune to eavesdropping. This technology promises to enhance data security in industries such as finance and defense.

Case Study: Quantum-Secure Banking Systems: Several financial institutions are exploring the use of QKD to secure transactions and communications, ensuring the integrity and confidentiality of sensitive financial data.

Diagram of Quantum Key Distribution:

Image: Diagram illustrating the process of Quantum Key Distribution, highlighting the secure exchange of cryptographic keys using quantum entanglement.

Drug Discovery: Quantum computing has the potential to significantly accelerate drug discovery by simulating molecular interactions at an unprecedented scale and accuracy.

- **Molecular Simulations:** Quantum computers can efficiently simulate the quantum states of complex molecules, enabling researchers to understand their properties and interactions more accurately.
- **Protein Folding:** Quantum algorithms can help predict protein folding, a critical aspect of understanding biological processes and developing new drugs.

Case Study: Pharmaceutical Research: Pharmaceutical companies are investing in quantum computing to accelerate the discovery of new drugs and optimize existing ones. For example, quantum simulations can help identify promising drug candidates and predict their behavior in the human body.

Materials Science: Quantum computing can transform materials science by enabling the discovery and design of new materials with tailored properties.

- **Material Properties:** Quantum simulations can predict the electronic, magnetic, and optical properties of materials, facilitating the development of new materials for various applications.
- **Nanotechnology:** Quantum computing can help design nanoscale materials with specific characteristics, leading to advancements in fields such as electronics, energy storage, and catalysis.

Case Study: Advanced Materials Development: Research institutions and companies are using quantum computing to explore new materials for use in semiconductors, batteries, and renewable energy technologies.

Hybrid Approaches

Integrating Quantum and Classical Computing: Hybrid quantum-classical approaches combine the strengths of quantum and classical computing to solve complex problems more efficiently. These approaches leverage classical computers for tasks they perform well and quantum computers for tasks that benefit from quantum parallelism and entanglement.

Quantum-Classical Algorithms: Several hybrid algorithms have been developed to take advantage of both quantum and classical resources.

- **Variational Quantum Eigensolver (VQE):** VQE is a hybrid algorithm used for finding the ground state energy of molecules. It combines quantum circuits to evaluate the energy and classical optimization techniques to adjust the parameters of the quantum circuits.
- **Quantum Approximate Optimization Algorithm (QAOA):** QAOA is used for solving combinatorial optimization problems. It employs a quantum-classical loop where the quantum computer evaluates the cost function, and the classical computer optimizes the parameters.

Case Study: Hybrid Quantum-Classical Systems: Companies like IBM and Google are developing hybrid quantum-classical systems to tackle problems in chemistry, optimization, and machine learning. These systems use classical computers to manage the overall process and quantum computers to perform specific quantum tasks.

Enhanced Problem-Solving Capabilities: Hybrid approaches can significantly enhance problem-solving capabilities across various fields.

- **Machine Learning:** Quantum machine learning algorithms can process and analyse large datasets more efficiently, leading to improved pattern recognition and predictive models.
- **Optimization Problems:** Hybrid algorithms can solve complex optimization problems in logistics, finance, and

manufacturing, providing more efficient and cost-effective solutions.

Future Prospects: As quantum computing technology continues to mature, hybrid approaches are expected to become more prevalent, offering powerful tools for tackling some of the most challenging problems in science, engineering, and industry.

This detailed exploration of the opportunities for innovation in quantum computing highlights the transformative potential of this technology across various industries. By leveraging hybrid quantum-classical approaches, researchers and practitioners can harness the strengths of both paradigms to solve complex problems and drive advancements in fields such as cryptography, drug discovery, materials science, and more.

1.5 Looking Ahead: The Future of Quantum Computing

1.5.1 Quantum Supremacy and Beyond

The concept of quantum supremacy represents a critical milestone in the development of quantum computing, marking the point at which a quantum computer can solve a problem that is infeasible for classical computers. This section delves into the definition and significance of quantum supremacy, and explores future milestones that are expected to shape the trajectory of quantum computing over the next decade and beyond.

Defining Quantum Supremacy

What is Quantum Supremacy? Quantum supremacy is the term used to describe the moment when a quantum computer performs a calculation that is beyond the capabilities of the most advanced classical supercomputers. This concept was first introduced by John Preskill in 2012 to highlight the practical demonstration of quantum computational advantage.

Significance of Quantum Supremacy:

- **Proof of Principle:** Demonstrating quantum supremacy provides tangible proof that quantum computers can outperform classical computers for specific tasks, validating decades of theoretical research and development.
- **Benchmark for Progress:** Quantum supremacy serves as a benchmark for measuring the progress of quantum computing technology. Achieving this milestone signals a significant leap forward in computational capabilities.
- **Catalyst for Innovation:** Reaching quantum supremacy stimulates further research and investment in quantum computing, driving innovation and the development of new algorithms and applications.

Google's Quantum Supremacy Claim (2019): In October 2019, Google announced that its 53-qubit quantum processor, Sycamore, had achieved quantum supremacy. The Sycamore processor completed a specific computation in 200 seconds that would have taken the world's fastest supercomputer approximately 10,000 years to solve.

Diagram of Quantum Supremacy:

Image: Diagram illustrating Google's Sycamore processor achieving quantum supremacy by outperforming classical supercomputers on a specific task.

Future Milestones

Beyond Quantum Supremacy: While achieving quantum supremacy is a significant milestone, the journey of quantum computing is far from over. The next phase involves developing practical, scalable, and fault-tolerant quantum computers that can tackle real-world problems across various industries.

Key Predictions for Future Achievements:

1. Fault-Tolerant Quantum Computing: Developing error-correcting codes and fault-tolerant architectures is crucial for building reliable quantum computers. Future milestones in this area include:

- **Quantum Error Correction:** Implementation of efficient quantum error-correcting codes, such as surface codes, to protect quantum information from errors and decoherence.
- **Logical Qubits:** Creating logical qubits from multiple physical qubits to perform stable and error-free quantum computations.

Example of Fault-Tolerant Quantum Computing: A fault-tolerant quantum computer using surface codes can perform complex computations with high accuracy, despite the presence of errors in individual qubits.

2. Quantum Advantage in Practical Applications: Quantum advantage refers to the ability of quantum computers to provide a significant speedup or performance improvement for practical applications compared to classical computers. Future milestones include:

- **Optimization Problems:** Quantum computers solving complex optimization problems in logistics, finance, and manufacturing more efficiently than classical algorithms.
- **Quantum Chemistry:** Simulating complex chemical reactions and materials with unprecedented accuracy, leading to breakthroughs in drug discovery and materials science.

Case Study: Quantum Optimization: A quantum computer optimizing supply chain logistics can reduce costs and increase efficiency by solving large-scale combinatorial problems faster than classical methods.

3. Quantum Network and Communication: The development of quantum networks and communication systems will enable secure and high-speed transmission of quantum information over long distances. Future milestones include:

- **Quantum Internet:** Establishing a global quantum network that connects quantum computers, sensors, and communication devices, enabling new applications in secure communication and distributed quantum computing.
- **Quantum Repeaters:** Developing quantum repeaters to extend the range of quantum communication by overcoming losses in optical fibers.

Example of Quantum Internet: A global quantum network allows for secure communication and distributed quantum computing, enabling new forms of collaboration and data sharing.

4. Integration with Classical Computing: Hybrid quantum-classical computing systems that integrate quantum processors with classical computers will become more prevalent, leveraging the strengths of both paradigms for enhanced problem-solving capabilities. Future milestones include:

- **Quantum Accelerators:** Development of quantum accelerators that can be integrated with classical supercomputers to speed up specific tasks.
- **Quantum Machine Learning:** Application of quantum algorithms to enhance machine learning models, leading to improved performance in data analysis and pattern recognition.

Case Study: Quantum Machine Learning: A hybrid quantum-classical system using quantum algorithms to train machine learning models can achieve faster and more accurate results in fields such as image recognition and natural language processing.

This detailed exploration of quantum supremacy and future milestones highlights the transformative potential of quantum computing and the path forward beyond achieving quantum supremacy. By focusing on fault-tolerant computing, practical applications, quantum networks, and hybrid systems, researchers and developers can unlock new opportunities and drive advancements that will reshape industries and everyday life over the next decade and beyond.

1.5.2 Quantum Ecosystem

The growth of the quantum computing industry relies heavily on a robust ecosystem composed of startups, developers, the open-source community, and a well-trained workforce. This section delves into the roles of these key players and emphasizes the importance of education and training in building a skilled quantum workforce.

Building the Ecosystem

Startups: Startups play a crucial role in driving innovation in the quantum computing space. They often focus on niche areas, developing new technologies and applications that push the boundaries of what is possible with quantum computing.

- **Pioneering Technologies:** Startups are at the forefront of developing cutting-edge quantum hardware and software solutions. For example, companies like IonQ and Rigetti Computing are making significant strides in quantum hardware development, while others like Zapata Computing focus on quantum software and algorithms.
- **Agility and Innovation:** Startups have the agility to experiment with new ideas and approaches, often leading to rapid advancements and breakthroughs in the field.

Case Study: IonQ IonQ is a startup that has developed a unique trapped-ion quantum computer. Their technology is known for high-fidelity operations and modularity, making it a promising platform for scalable quantum computing.

Developers: Developers are the backbone of the quantum computing ecosystem, creating the software and algorithms that enable practical applications of quantum technology.

- **Quantum Programming:** Developers are building quantum software using programming languages and frameworks like Qiskit, Cirq, and PyQuil. These tools allow for the creation and testing of quantum algorithms on simulators and actual quantum hardware.
- **Algorithm Development:** Quantum developers are responsible for designing and optimizing algorithms that leverage the unique capabilities of quantum computing to solve specific problems in fields such as cryptography, chemistry, and machine learning.

Case Study: Qiskit Developer Community Qiskit, developed by IBM, has a vibrant open-source community where developers contribute to the growth of quantum software. The community collaborates on projects, shares knowledge, and helps newcomers learn quantum programming.

Open-Source Community: The open-source community is vital for the democratization and rapid advancement of quantum technologies. Open-source projects provide accessible resources for learning, experimenting, and contributing to the development of quantum computing.

- **Collaboration and Sharing:** Open-source platforms encourage collaboration among researchers, developers, and enthusiasts from around the world, accelerating the pace of innovation.
- **Accessible Resources:** Open-source projects make quantum computing tools and libraries accessible to a broader audience, fostering inclusivity and encouraging diverse contributions.

Example: Open-Source Quantum Projects Projects like Qiskit, Cirq, and ProjectQ provide open-source quantum computing frameworks that enable users to build and test quantum algorithms. These platforms often include extensive documentation, tutorials, and community support to help users get started.

Educational and Workforce Development

Importance of Education: Building a skilled quantum workforce is critical for the sustained growth and development of the quantum computing industry. Educational programs and training initiatives are essential to equip the next generation of scientists, engineers, and developers with the knowledge and skills needed to advance quantum technologies.

- **University Programs:** Many universities are now offering specialized programs in quantum computing, quantum information science, and related fields. These programs provide foundational knowledge and hands-on experience with quantum systems.
- **Online Courses and Certifications:** Online platforms like edX, Coursera, and Udacity offer courses and certifications in quantum computing, making education more accessible to a global audience.

Example: MIT xPro Quantum Computing Program MIT xPro offers an online professional certificate in quantum computing, covering topics such as quantum mechanics, quantum algorithms, and quantum hardware. The program is designed for professionals looking to upskill and transition into the quantum computing field.

Training Initiatives: In addition to formal education, training initiatives and workshops play a crucial role in workforce development. These initiatives provide practical experience and opportunities for hands-on learning.

- **Hackathons and Competitions:** Quantum computing hackathons and competitions challenge participants to

solve real-world problems using quantum algorithms, fostering innovation and skill development.
- **Industry Partnerships:** Collaborations between academia and industry provide students and professionals with internships, research opportunities, and exposure to cutting-edge quantum technologies.

Example: IBM Quantum Challenge IBM regularly hosts the IBM Quantum Challenge, a global competition where participants solve complex problems using Qiskit. These challenges help participants deepen their understanding of quantum computing and develop practical skills.

Building a Skilled Workforce: The development of a skilled quantum workforce involves a multi-faceted approach, including:

- **Curriculum Development:** Integrating quantum computing topics into existing science, technology, engineering, and mathematics (STEM) curricula at all educational levels.
- **Professional Development:** Providing ongoing training and professional development opportunities for current scientists, engineers, and IT professionals to transition into the quantum field.
- **Community Engagement:** Encouraging community engagement and public awareness of quantum computing through outreach programs, public lectures, and science festivals.

Future Prospects: As the quantum computing industry continues to grow, the demand for skilled professionals will increase. By investing in education and workforce development, we can ensure a steady supply of talent to drive the next wave of quantum innovation.

This detailed exploration of the quantum ecosystem emphasizes the critical roles played by startups, developers, the open-source community, and educational institutions in advancing quantum technology. By fostering innovation and building a skilled workforce, we can harness the full potential of quantum computing to transform industries and reshape everyday life over the next decade and beyond.

1.5.3 Societal Impact

The advent of quantum computing brings with it a range of societal implications that extend beyond technological advancements. As quantum technologies develop, it is crucial to consider their ethical implications and establish appropriate regulatory frameworks to ensure their responsible and beneficial deployment. This section explores the ethical considerations and potential policy and regulatory approaches for guiding the development of quantum computing.

Ethical Considerations

Privacy and Security: Quantum computing has the potential to revolutionize data security and privacy, but it also poses significant challenges.

- **Breaking Encryption:** Quantum computers, particularly through Shor's algorithm, could break widely used

cryptographic schemes like RSA and ECC, posing a threat to data security across the internet.
- **Quantum-Resistant Cryptography:** The development and deployment of quantum-resistant cryptographic methods are essential to safeguard data against quantum threats. Organizations must transition to these new cryptographic standards to maintain data privacy and security.

Case Study: Securing Communication Networks Governments and companies are investing in quantum key distribution (QKD) to protect sensitive communication networks from potential quantum threats. QKD leverages quantum mechanics to enable secure key exchange that is theoretically immune to eavesdropping.

Equity and Accessibility: Ensuring that the benefits of quantum computing are equitably distributed and accessible is a significant ethical consideration.

- **Digital Divide:** The high cost and complexity of quantum technologies could exacerbate the digital divide, making advanced computing resources accessible only to wealthy nations and large corporations.
- **Inclusive Development:** Efforts should be made to promote inclusive development and provide access to quantum computing resources for educational institutions, startups, and developing countries.

Example: Quantum Computing for Education Initiatives like IBM's Quantum Experience provide free access to quantum computing resources for students, researchers, and educators worldwide, promoting inclusivity and democratizing access to advanced technologies.

Ethical Use of Quantum Technologies: The ethical use of quantum technologies involves considering their broader impact on society and ensuring they are used for the common good.

- **Ethical Frameworks:** Developing ethical frameworks and guidelines for the use of quantum computing in various applications, such as AI and data analysis, to prevent misuse and protect individual rights.
- **Social Impact Assessments:** Conducting social impact assessments to evaluate the potential effects of quantum technologies on different societal groups and addressing any adverse outcomes.

Case Study: AI and Quantum Computing Combining AI with quantum computing has the potential to enhance decision-making processes. However, it is essential to ensure that these technologies are used ethically, with considerations for fairness, transparency, and accountability.

Policy and Regulation

Regulatory Frameworks: Establishing robust regulatory frameworks is crucial for guiding the responsible development and deployment of quantum computing technologies.

- **International Cooperation:** Quantum computing is a global endeavor, and international cooperation is essential for developing standardized regulations and policies. Organizations like the International Telecommunication Union (ITU) and the World Economic Forum (WEF) can play key roles in fostering global collaboration.
- **National Strategies:** Countries should develop national quantum strategies that outline their approach to quantum research, development, and deployment, including funding priorities and ethical guidelines.

Example: National Quantum Initiatives Countries like the United States, China, and members of the European Union have launched national quantum initiatives to coordinate research efforts, allocate funding, and develop regulatory policies for quantum technologies.

Privacy and Data Protection: Regulations must address the unique privacy and data protection challenges posed by quantum computing.

- **Data Encryption:** Updating data encryption standards to be quantum-resistant is essential to protect sensitive information from future quantum attacks.
- **Privacy Laws:** Ensuring that privacy laws keep pace with technological advancements in quantum computing, providing clear guidelines on data handling and protection.

Case Study: GDPR and Quantum Computing The General Data Protection Regulation (GDPR) in the European Union provides a framework for data protection that could be adapted to include provisions for quantum computing, ensuring that personal data remains secure.

Ethical AI and Quantum Computing: Policies should also address the ethical use of AI in conjunction with quantum computing.

- **Transparency and Accountability:** Ensuring transparency in how quantum-enhanced AI systems make decisions and holding developers accountable for their algorithms' outcomes.
- **Bias Mitigation:** Developing standards and practices to identify and mitigate biases in quantum-enhanced AI systems, promoting fairness and equity.

Example: AI Ethics Guidelines Organizations like the European Commission have published guidelines for ethical AI that could be expanded to include considerations for quantum-enhanced AI systems.

Future-Proofing Regulations: As quantum technologies continue to evolve, regulatory frameworks must be adaptable to address emerging challenges and opportunities.

- **Dynamic Policies:** Developing dynamic and flexible policies that can be updated as new quantum technologies and applications emerge.
- **Stakeholder Engagement:** Engaging a broad range of stakeholders, including technologists, ethicists, policymakers, and the public, in the regulatory process to ensure comprehensive and inclusive policy development.

Case Study: Adaptive Quantum Policies Countries adopting adaptive regulatory approaches can better manage the rapid advancements in quantum technology, ensuring that policies remain relevant and effective.

This detailed exploration of the societal impact of quantum computing emphasizes the importance of ethical considerations and robust regulatory frameworks in guiding the responsible development and deployment of quantum technologies. By addressing privacy and security concerns, promoting equitable access, and ensuring ethical use, we can harness the transformative potential of quantum computing for the benefit of all.

Chapter 2: The AI Revolution

2.1 Overview of Artificial Intelligence

2.1.1 Defining AI

Artificial Intelligence (AI) is a transformative technology that has already begun reshaping various aspects of our lives, from personal assistants and recommendation systems to advanced robotics and autonomous vehicles. This section provides a comprehensive definition of AI, explores the different types of AI, and traces the historical development of this fascinating field.

What is AI?

Comprehensive Definition of AI: Artificial Intelligence is a branch of computer science that aims to create machines capable of performing tasks that typically require human intelligence. These tasks include learning from experience, understanding natural language, recognizing patterns, solving problems, and making decisions.

- **Learning:** AI systems can learn from data and improve their performance over time without being explicitly programmed. This capability is primarily driven by machine learning algorithms.
- **Reasoning:** AI systems can process information and make logical inferences to solve complex problems, often employing techniques from symbolic reasoning and logic.
- **Perception:** AI can interpret and understand sensory data, such as visual images and audio signals, enabling applications like image recognition and speech processing.
- **Natural Language Processing (NLP):** AI systems can understand and generate human language, facilitating communication between humans and machines.

Example: AI-powered virtual assistants, like Siri and Alexa, use natural language processing to understand user queries and machine learning to improve their responses over time.

Types of AI

Narrow AI: Narrow AI, also known as weak AI, is designed to perform a specific task or a narrow range of tasks. These systems operate under a limited set of constraints and do not possess general intelligence.

- **Examples:** Voice assistants like Siri, image recognition systems, recommendation engines, and autonomous vehicles are all examples of narrow AI.

General AI: General AI, or strong AI, refers to systems that possess the ability to understand, learn, and apply knowledge across a wide range of tasks, similar to human intelligence. General AI can reason, plan, and solve problems in diverse domains without being specifically programmed for each task.

- **Current Status:** General AI remains a theoretical concept and has not yet been achieved. Research in this area focuses on developing more adaptable and flexible AI systems.

Superintelligent AI: Superintelligent AI is an advanced form of AI that surpasses human intelligence across all domains. It would be capable of performing tasks and making decisions that are beyond human cognitive capabilities.

- **Implications:** The development of superintelligent AI poses significant ethical and existential risks. Ensuring that such AI systems align with human values and objectives is a critical area of research.

Historical Context

Early Concepts: The concept of artificial intelligence dates back to ancient myths and stories about artificial beings endowed with intelligence. However, the formal study of AI began in the mid-20th century.

- **Alan Turing:** In 1950, British mathematician Alan Turing published a seminal paper titled "Computing Machinery and

Intelligence," which introduced the idea of a machine that could mimic human intelligence. Turing proposed the Turing Test as a criterion for determining whether a machine can exhibit intelligent behavior indistinguishable from that of a human.

The Birth of AI: The field of AI was formally established in 1956 at the Dartmouth Conference, organized by John McCarthy, Marvin Minsky, Nathaniel Rochester, and Claude Shannon. This event is considered the birth of AI as a scientific discipline.

- **Early AI Programs:** Early AI research focused on symbolic reasoning and problem-solving. Notable programs from this era include the Logic Theorist (1955) and General Problem Solver (1957), which were designed to simulate human problem-solving processes.

AI Winters: The history of AI has seen periods of high optimism and significant funding, followed by periods of disappointment and reduced investment, known as AI winters. These cycles were often driven by unrealistic expectations and the slow progress of AI research.

- **First AI Winter (1974-1980):** Funding cuts and skepticism about the feasibility of AI led to a period of reduced research activity.
- **Second AI Winter (1987-1993):** The collapse of the AI market, particularly for expert systems, resulted in another period of decreased funding and interest.

Modern Advancements: The resurgence of AI in the 21st century has been driven by breakthroughs in machine learning, the availability of large datasets, and advances in computational power.

- **Deep Learning:** The development of deep learning algorithms, which use artificial neural networks to model complex patterns in data, has revolutionized fields such as image and speech recognition.

- **AI Applications:** Modern AI applications span various domains, including healthcare (e.g., diagnostic systems), finance (e.g., algorithmic trading), and entertainment (e.g., recommendation systems).

Case Study: AlphaGo In 2016, Google's DeepMind developed AlphaGo, an AI system that defeated the world champion Go player. This achievement demonstrated the potential of deep learning and reinforcement learning to solve complex problems.

This detailed exploration of AI provides a comprehensive understanding of its definition, types, and historical development. By examining the capabilities of AI, the different forms it can take, and its historical context, we gain insight into the transformative potential of AI and its role in shaping the future of technology and society.

2.1.2 Key Concepts in AI

Artificial Intelligence (AI) encompasses a variety of key concepts that enable machines to perform tasks requiring human-like intelligence. This section delves into three foundational areas of AI: machine learning, deep learning, and natural language processing (NLP). Each of these areas plays a crucial role in advancing AI technologies and their applications across various industries.

Machine Learning

Overview of Machine Learning (ML): Machine learning is a subset of AI that involves training algorithms to make predictions or decisions based on data. ML algorithms learn from historical data to identify patterns and make informed predictions on new data. There are three main types of machine learning: supervised learning, unsupervised learning, and reinforcement learning.

- **Supervised Learning:** In supervised learning, algorithms are trained on labeled data, meaning that each training example is paired with an output label. The model learns to map inputs to the correct outputs based on this training data.
    - **Examples:**
        - **Classification:** Identifying whether an email is spam or not.
        - **Regression:** Predicting house prices based on features like location, size, and number of bedrooms.
    - **Popular Algorithms:**
        - **Linear Regression:** Used for regression tasks.
        - **Support Vector Machines (SVM):** Used for classification tasks.
        - **Decision Trees:** Used for both classification and regression tasks.
- **Unsupervised Learning:** Unsupervised learning algorithms are trained on unlabeled data, meaning the data does not have explicit outputs. The goal is to identify underlying patterns or structures in the data.
    - **Examples:**
        - **Clustering:** Grouping customers based on purchasing behavior.
        - **Dimensionality Reduction:** Reducing the number of features in a dataset while preserving important information.

- **Popular Algorithms:**
  - **K-Means Clustering:** Groups data into k number of clusters.
  - **Principal Component Analysis (PCA):** Reduces the dimensionality of data.

- **Reinforcement Learning:** In reinforcement learning, an agent learns to make decisions by interacting with an environment. The agent receives rewards or penalties based on its actions and aims to maximize the cumulative reward over time.

  - **Examples:**
    - **Game Playing:** Training an AI to play chess or Go.
    - **Robotics:** Teaching a robot to navigate a maze.

  - **Popular Algorithms:**
    - **Q-Learning:** A value-based reinforcement learning algorithm.
    - **Policy Gradient Methods:** Learn policies that directly map states to actions.

Deep Learning

Introduction to Deep Learning: Deep learning is a specialized subset of machine learning that uses neural networks with many layers (deep neural networks) to model complex patterns in data. Deep learning has driven significant advancements in fields such as computer vision, speech recognition, and natural language processing.

- **Neural Networks:** Neural networks are the foundation of deep learning. They consist of layers of interconnected nodes (neurons), each performing simple computations. The network learns by adjusting the weights of the connections based on the error of its predictions.

- **Feedforward Neural Networks:** The simplest type of neural network, where information flows in one direction from input to output.
- **Convolutional Neural Networks (CNNs):** Specialized for processing grid-like data, such as images. They use convolutional layers to automatically learn spatial hierarchies of features.
- **Recurrent Neural Networks (RNNs):** Designed for sequential data, such as time series or text. They have loops that allow information to persist across steps.

- **Training Deep Neural Networks:** Deep learning models are trained using large datasets and high computational power. The training process involves forward propagation, where data passes through the network, and backpropagation, where the network updates its weights based on the error gradient.

- **Applications of Deep Learning:**
  - **Computer Vision:** Object detection, image classification, and facial recognition.
  - **Speech Recognition:** Transcribing spoken language into text.
  - **Natural Language Processing:** Language translation, sentiment analysis, and chatbots.

Case Study: Image Classification with CNNs Convolutional Neural Networks (CNNs) have revolutionized image classification by automatically learning hierarchical features from raw pixel data, leading to state-of-the-art performance on various benchmarks.

Natural Language Processing (NLP)

Explanation of NLP: Natural Language Processing (NLP) is a branch of AI focused on the interaction between computers and humans through natural language. NLP enables machines to understand, interpret, and generate human language.

- **Core Tasks in NLP:**
    - **Tokenization:** Splitting text into words or subwords.
    - **Part-of-Speech Tagging:** Identifying the grammatical parts of speech for each token.
    - **Named Entity Recognition (NER):** Identifying entities such as people, organizations, and locations in text.
    - **Sentiment Analysis:** Determining the sentiment or emotion expressed in text.

- **Advanced NLP Techniques:**
    - **Word Embeddings:** Representing words as vectors in a continuous vector space. Techniques like Word2Vec and GloVe capture semantic relationships between words.
    - **Transformers:** A deep learning architecture that uses self-attention mechanisms to process sequential data. Transformers have become the standard for many NLP tasks.

Applications of NLP:

- **Machine Translation:** Translating text from one language to another.
- **Chatbots and Virtual Assistants:** Automating customer service and providing conversational interfaces.
- **Text Summarization:** Automatically generating concise summaries of long documents.

Case Study: GPT-4 OpenAI's GPT-4 is a state-of-the-art language model that uses the transformer architecture to generate coherent and contextually relevant text. It has applications in content creation, code generation, and conversational AI.

This detailed exploration of key concepts in AI provides a comprehensive understanding of machine learning, deep learning, and natural language processing. By examining these foundational areas, we gain insight into the capabilities of AI technologies and their transformative potential across various industries.

2.1.3 AI vs. Human Intelligence

Artificial Intelligence (AI) has made significant strides in recent years, achieving capabilities that rival or even surpass human performance in specific tasks. However, the relationship between AI and human intelligence is complex, with each possessing unique strengths and limitations. This section explores the comparison between AI and human intelligence, highlighting their respective capabilities and limitations, and discusses how AI can augment human intelligence to create powerful synergies.

Capabilities and Limitations

Capabilities of AI: AI systems have demonstrated remarkable capabilities in various domains, leveraging vast computational power and advanced algorithms to process and analyse data at scales unattainable by humans.

- **Speed and Efficiency:** AI can perform calculations and process data much faster than humans. For example, AI algorithms can analyse large datasets in seconds, a task that would take humans significantly longer.

- **Pattern Recognition:** AI excels at recognizing patterns in data, enabling applications such as image recognition, natural language processing, and anomaly detection. Deep learning models, for instance, can identify objects in images with high accuracy.
- **Consistency and Precision:** AI systems can consistently perform tasks with high precision and without fatigue. This consistency is critical in applications like medical diagnosis and industrial automation, where errors can have significant consequences.
- **Learning from Data:** Machine learning algorithms enable AI to learn from vast amounts of data, improving their performance over time. This capability is evident in recommendation systems and personalized marketing, where AI adapts to user preferences.

Limitations of AI: Despite its capabilities, AI has several limitations that distinguish it from human intelligence.

- **Lack of Generalization:** AI systems are typically designed for specific tasks and lack the generalization capabilities of human intelligence. An AI trained for image recognition, for example, cannot easily transfer its knowledge to a different domain, such as natural language processing.
- **Absence of Common Sense:** AI lacks common sense reasoning and the ability to understand context as humans do. This limitation can lead to incorrect or nonsensical outputs in complex or ambiguous situations.
- **Ethical and Bias Issues:** AI systems can inherit biases from the data they are trained on, leading to unfair or discriminatory outcomes. Addressing these biases and ensuring ethical AI deployment is a significant challenge.
- **Dependence on Data:** AI performance heavily relies on the quality and quantity of training data. Inadequate or biased data can result in poor model performance and unintended consequences.

Example: AI in Healthcare AI has revolutionized healthcare with applications such as diagnostic imaging and personalized treatment plans. However, AI models must be carefully trained and validated to ensure accuracy and avoid biases that could affect patient outcomes.

.

Capabilities of Human Intelligence: Human intelligence encompasses a broad range of cognitive abilities that enable flexible and adaptive problem-solving.

- **Generalization:** Humans can generalize knowledge across different domains, applying insights from one area to solve problems in another.
- **Common Sense Reasoning:** Humans possess common sense and can understand and interpret context, allowing for nuanced decision-making.
- **Creativity and Innovation:** Human intelligence fosters creativity and the ability to generate novel ideas and solutions. This capability is crucial in fields such as art, science, and entrepreneurship.
- **Emotional Intelligence:** Humans can understand and manage emotions, enabling effective communication, empathy, and social interactions.

Limitations of Human Intelligence: While human intelligence is versatile, it also has inherent limitations.

- **Cognitive Biases:** Humans are susceptible to cognitive biases that can affect decision-making and judgment.
- **Limited Processing Speed:** Human brains have limited processing speed and capacity compared to AI systems, which can handle vast amounts of data rapidly.
- **Fatigue and Inconsistency:** Humans are prone to fatigue and inconsistencies in performance, particularly in repetitive or monotonous tasks.

Example: Human Creativity in Art Human creativity is exemplified in the arts, where artists use their imagination and emotions to create unique and expressive works. AI can assist in the creative process but cannot fully replicate the depth of human artistic expression.

Complementary Roles

Augmenting Human Intelligence: AI and human intelligence can complement each other, creating synergies that enhance overall capabilities and enable new possibilities.

- **Human-AI Collaboration:** By combining the strengths of AI and human intelligence, we can achieve better outcomes in various fields. For instance, AI can handle data-intensive tasks, allowing humans to focus on strategic and creative aspects.
- **Decision Support Systems:** AI can provide valuable insights and recommendations, assisting humans in making informed decisions. This approach is widely used in finance, healthcare, and business management.
- **Enhanced Productivity:** AI-powered tools and automation can increase productivity by performing routine tasks, freeing up human resources for more complex and value-added activities.

Case Study: AI in Financial Services In the financial sector, AI algorithms analyse market trends and generate trading signals, while human traders use their expertise to make final investment decisions. This collaboration enhances the accuracy and efficiency of financial operations.

Future Prospects: The future of AI and human intelligence lies in their integration, with AI augmenting human capabilities and vice versa. This synergy has the potential to drive innovation, improve quality of life, and address complex global challenges.

- **Education and Training:** AI can personalize education and training, tailoring content to individual learning styles and needs, thus enhancing human potential.
- **Healthcare and Medicine:** AI can assist healthcare professionals in diagnosing diseases, developing treatment plans, and managing patient care, ultimately improving health outcomes.
- **Environmental Sustainability:** AI can analyse environmental data and develop strategies for sustainable resource management, helping address climate change and environmental degradation.

Example: Personalized Learning with AI AI-powered educational platforms use machine learning to adapt lessons and assessments to each student's progress and learning style, providing a more effective and personalized learning experience.

This detailed exploration of AI versus human intelligence highlights the unique capabilities and limitations of each and underscores the potential for their complementary roles. By leveraging the strengths of both AI and human intelligence, we can create powerful synergies that enhance our abilities and address complex challenges, driving progress and innovation in the coming decades.

2.2 Core Components and Techniques in AI

2.2.1 Algorithms and Models

Understanding the algorithms and models that underpin artificial intelligence (AI) is essential for grasping how AI systems function and how they can be applied to various tasks. This section explores common algorithms used in machine learning and introduces key deep learning models, providing a comprehensive overview of the techniques that drive modern AI.

Common Algorithms

Decision Trees: Decision trees are a popular machine learning algorithm used for classification and regression tasks. They model decisions and their possible consequences as a tree structure, where each internal node represents a "decision" based on an attribute, each branch represents the outcome of that decision, and each leaf node represents a final classification or output.

- **How They Work:**
    - **Splitting:** The data is split into subsets based on the value of an attribute. This process is repeated recursively on each subset, creating branches of the tree.
    - **Decision Nodes:** Nodes where decisions are made based on attribute values.
    - **Leaf Nodes:** Terminal nodes that provide the final output (classification or regression value).

- **Advantages:**
    - **Easy to Understand and Interpret:** Decision trees are intuitive and can be visualized, making them easy to interpret.
    - **Handling Both Numerical and Categorical Data:** They can handle various types of data without requiring normalization.

- **Disadvantages:**
    - **Prone to Overfitting:** Decision trees can overfit the training data, especially if they are deep with many nodes.

Support Vector Machines (SVM): Support vector machines are supervised learning models used for classification and regression tasks. SVMs work by finding the hyperplane that best separates the data points of different classes in a high-dimensional space.

- **How They Work:**
    - **Hyperplane:** The algorithm identifies the hyperplane that maximizes the margin between the closest data points of different classes (support vectors).
    - **Kernel Trick:** SVMs can efficiently perform a non-linear classification using the kernel trick, which implicitly maps the input features into high-dimensional space.

- **Advantages:**
    - **Effective in High-Dimensional Spaces:** SVMs are effective in cases where the number of dimensions exceeds the number of samples.
    - **Robust to Overfitting:** Especially in high-dimensional spaces, SVMs are less prone to overfitting.

- **Disadvantages:**
    - **Computationally Intensive:** Training an SVM can be resource-intensive, especially with large datasets.

K-Nearest Neighbors (KNN): K-nearest neighbors is a simple, instance-based learning algorithm used for classification and regression tasks. KNN classifies a data point based on the majority class among its k-nearest neighbors in the feature space.

- **How They Work:**

- **Distance Metrics:** Common distance metrics include Euclidean distance, Manhattan distance, and Minkowski distance.
- **Majority Voting:** The algorithm assigns the class of the majority of the k-nearest neighbors to the new data point.

- **Advantages:**
  - **Simple and Intuitive:** KNN is easy to implement and understand.
  - **No Training Phase:** It does not have a training phase, making it fast for small datasets.

- **Disadvantages:**
  - **Computationally Expensive:** The algorithm needs to compute the distance between the input and all training data points, which can be slow with large datasets.
  - **Sensitive to Irrelevant Features:** Performance can be affected by irrelevant features, and feature scaling can significantly impact results.

Deep Learning Models

Convolutional Neural Networks (CNNs): Convolutional neural networks are a class of deep learning models primarily used for processing grid-like data such as images. CNNs are known for their ability to automatically and adaptively learn spatial hierarchies of features through backpropagation.

- **Key Components:**
  - **Convolutional Layers:** Apply convolution operations to the input, extracting local features.

- **Pooling Layers:** Reduce the dimensionality of the feature maps while retaining important information, typically using max pooling.
- **Fully Connected Layers:** Flatten the feature maps and connect them to output layers for classification or regression.

- **Applications:**
  - **Image Classification:** Identifying objects within images.
  - **Object Detection:** Locating and classifying multiple objects within an image.
  - **Image Segmentation:** Partitioning an image into meaningful segments.

Recurrent Neural Networks (RNNs): Recurrent neural networks are designed for sequential data, such as time series or natural language. RNNs have connections that form cycles, allowing information to persist across time steps.

- **Key Components:**
  - **Recurrent Layers:** Contain loops that allow information to be passed from one time step to the next.
  - **Long Short-Term Memory (LSTM) and Gated Recurrent Unit (GRU):** Variants of RNNs that address the vanishing gradient problem and can capture long-term dependencies.

- **Applications:**
  - **Language Modeling and Text Generation:** Predicting the next word in a sequence.

- **Speech Recognition:** Transcribing spoken language into text.
- **Time Series Prediction:** Forecasting future values based on past observations.

Generative Adversarial Networks (GANs): Generative adversarial networks consist of two neural networks, a generator and a discriminator, that are trained together in a process known as adversarial training. GANs are used to generate new, synthetic data that resembles the training data.

- **Key Components:**
    - **Generator:** Creates synthetic data samples from random noise.
    - **Discriminator:** Evaluates whether a given data sample is real (from the training data) or fake (generated by the generator).

- **Training Process:**
    - The generator tries to create data that is indistinguishable from real data.
    - The discriminator tries to correctly classify data as real or fake.
    - Both networks improve through a competitive process until the generator produces realistic data.

- **Applications:**
    - **Image Generation:** Creating realistic images from random noise.
    - **Data Augmentation:** Generating additional training data to improve model performance.
    - **Art and Design:** Generating artwork and creative designs.

This detailed exploration of algorithms and models in AI provides a comprehensive understanding of the key techniques that drive machine learning and deep learning. By examining common algorithms such as decision trees, support vector machines, and k-nearest neighbors, as well as deep learning models like convolutional neural networks, recurrent neural networks, and generative adversarial networks, we gain insight into the foundations of AI and their diverse applications.

2.2.2 Data and Training

The efficacy and performance of artificial intelligence (AI) models are heavily dependent on the quality and quantity of data used during their training. Understanding the role of data and the processes involved in training AI models is crucial for developing effective and reliable AI systems. This section delves into the importance of data in AI, outlines the training processes, and explains key steps such as data preprocessing, feature selection, and model evaluation.

Importance of Data

Role of Data in Training AI Models: Data is the cornerstone of AI, providing the raw material from which models learn patterns, make predictions, and derive insights. The quality and relevance of the data directly influence the performance of AI models.

- **Learning Patterns:** AI models, particularly those based on machine learning and deep learning, rely on data to identify underlying patterns and relationships. These patterns enable the models to generalize and make accurate predictions on new, unseen data.
- **Improving Accuracy:** High-quality data helps improve the accuracy and robustness of AI models. Clean, well-labeled, and representative datasets ensure that models can learn effectively and avoid overfitting or underfitting.

- **Diversity and Volume:** Large and diverse datasets provide a comprehensive view of the problem space, allowing models to capture a wide range of scenarios and variations. This diversity is crucial for building models that perform well across different contexts and environments.

Example: Training a Facial Recognition System A facial recognition system requires a large dataset of diverse facial images to learn how to accurately identify individuals. The dataset should include images with varying lighting conditions, angles, and facial expressions to ensure the model's robustness and accuracy.

Training Processes

Data Preprocessing: Data preprocessing is a crucial step in the AI training process, involving the transformation of raw data into a clean and usable format. This step ensures that the data is suitable for training the model.

- **Data Cleaning:** Removing noise, errors, and inconsistencies from the data. This includes handling missing values, correcting errors, and removing duplicates.
- **Normalization and Scaling:** Adjusting the data to a common scale without distorting differences in the ranges of values. Techniques such as min-max scaling and z-score normalization are commonly used.
- **Data Transformation:** Converting data into a format that is compatible with the model. This may include encoding categorical variables, creating dummy variables, and extracting features.

Example: Preprocessing Text Data for NLP In natural language processing (NLP), preprocessing text data involves steps such as tokenization (splitting text into words), stopword removal (removing common words like "the" and "is"), and stemming or lemmatization (reducing words to their root forms).

Feature Selection: Feature selection involves identifying the most relevant features in the dataset that contribute to the model's predictions. This step helps improve model performance by reducing dimensionality and removing irrelevant or redundant features.

- **Techniques for Feature Selection:**
    - **Filter Methods:** Use statistical techniques to evaluate the importance of features based on their relationship with the target variable. Examples include correlation coefficients and chi-square tests.
    - **Wrapper Methods:** Use a search algorithm to evaluate different subsets of features and select the combination that yields the best model performance. Examples include forward selection, backward elimination, and recursive feature elimination.
    - **Embedded Methods:** Perform feature selection during the model training process. Examples include regularization techniques like Lasso (L1) and Ridge (L2) regression.

Example: Feature Selection for a House Price Prediction Model
In a house price prediction model, relevant features might include the number of bedrooms, square footage, location, and age of the property. Irrelevant features, such as the owner's name or color of the house, should be excluded to improve model performance.

Training the Model: Training an AI model involves using the preprocessed data and selected features to learn the patterns and relationships within the data. This process typically involves splitting the data into training and validation sets, iteratively adjusting model parameters, and evaluating performance.

- **Training and Validation Split:** Dividing the dataset into separate training and validation sets. The training set is used to train the model, while the validation set is used to evaluate its performance.

- **Model Training:** The model learns by adjusting its parameters to minimize the error between its predictions and the actual outcomes. This process is iterative and involves techniques such as gradient descent.
- **Hyperparameter Tuning:** Optimizing the hyperparameters (e.g., learning rate, batch size) to improve model performance. Techniques like grid search and random search are commonly used.

Example: Training a Decision Tree Classifier In training a decision tree classifier, the model iteratively splits the data based on the most informative features, adjusts the tree structure to minimize classification error, and tunes hyperparameters such as tree depth and minimum samples per leaf.

Model Evaluation: Evaluating the trained model is essential to ensure its effectiveness and generalizability. This step involves using performance metrics to assess the model's accuracy, precision, recall, and other relevant measures.

- **Evaluation Metrics:**
    - **Accuracy:** The proportion of correctly predicted instances out of the total instances.
    - **Precision and Recall:** Metrics used to evaluate the performance of classification models, especially in imbalanced datasets.
    - **F1 Score:** The harmonic mean of precision and recall, providing a balanced measure of model performance.
    - **Mean Squared Error (MSE) and Root Mean Squared Error (RMSE):** Metrics used to evaluate the performance of regression models.
- **Cross-Validation:** A technique to assess model performance by splitting the data into multiple folds and training the

model on each fold. This helps ensure the model's robustness and reduces the risk of overfitting.

Example: Evaluating a Spam Detection Model A spam detection model can be evaluated using precision and recall to ensure it accurately identifies spam emails (precision) while minimizing false negatives (recall).

This detailed exploration of data and training processes in AI highlights the critical steps involved in preparing data, selecting features, training models, and evaluating their performance. By understanding these processes, we gain insight into how AI models are developed and optimized, ensuring their effectiveness and reliability in real-world applications.

2.2.3 AI Infrastructure

The development and deployment of artificial intelligence (AI) models require robust infrastructure, including specialized hardware and software frameworks. This section explores the hardware requirements essential for AI, such as GPUs and TPUs, and introduces popular AI frameworks and libraries that facilitate the creation and training of AI models.

Hardware Requirements

Overview of Hardware Needed for AI: AI workloads, especially those involving deep learning, demand significant computational power. Traditional CPUs, while powerful, are often insufficient for the parallel processing needs of AI. As a result, specialized hardware such as GPUs and TPUs has become essential for efficient AI processing.

- **Graphics Processing Units (GPUs):** GPUs are highly effective for AI tasks due to their ability to perform parallel computations. They are designed to handle multiple operations simultaneously, making them ideal for training deep neural networks.

- **Advantages:**
  - **Parallel Processing:** GPUs can perform thousands of parallel operations, significantly speeding up the training of AI models.
  - **High Throughput:** With many cores, GPUs provide the high throughput necessary for processing large datasets.
  - **Flexibility:** GPUs are versatile and can be used for a wide range of AI applications, from image processing to natural language processing.
- **Popular GPUs for AI:**
  - **NVIDIA Tesla and A100:** Designed specifically for AI and deep learning workloads, offering high performance and efficiency.
  - **AMD Radeon Instinct:** Another powerful option for AI and deep learning applications.

- **Tensor Processing Units (TPUs):** TPUs are specialized hardware accelerators developed by Google specifically for AI and machine learning tasks. They are designed to accelerate the training and inference of AI models, particularly deep learning models.

- **Advantages:**
  - **Optimized for TensorFlow:** TPUs are highly optimized for TensorFlow, one of the most popular AI frameworks.
  - **High Performance:** TPUs provide significant speedups for both training and inference, especially for large-scale models.
  - **Energy Efficiency:** TPUs are designed to be energy-efficient, reducing the cost and environmental impact of AI workloads.
- **Generations of TPUs:**
  - **TPU v1:** Focused on inference tasks.
  - **TPU v2 and v3:** Designed for both training and inference, offering increased performance and scalability.

Other Specialized Hardware:

- **Field-Programmable Gate Arrays (FPGAs):** FPGAs offer customizable hardware solutions for AI workloads. They provide flexibility and can be reconfigured for different tasks, making them suitable for specific AI applications.

- **Application-Specific Integrated Circuits (ASICs):** ASICs are custom-designed chips optimized for specific tasks. While they lack the flexibility of FPGAs, they offer high performance and efficiency for targeted AI applications.

Software Frameworks

Introduction to Popular AI Frameworks and Libraries: AI frameworks and libraries provide the tools and interfaces necessary for developing, training, and deploying AI models. These frameworks abstract the complexities of underlying computations, making it easier for developers and researchers to build sophisticated AI applications.

- **TensorFlow:** TensorFlow is an open-source framework developed by Google. It is widely used for both research and production AI applications and supports a range of tasks, including deep learning, machine learning, and neural network training.

    - **Features:**

        - **Flexibility:** TensorFlow supports various APIs and tools, allowing users to build and train models in different ways.

        - **Scalability:** TensorFlow can scale from individual CPUs and GPUs to large clusters of machines.

        - **Ecosystem:** TensorFlow has a rich ecosystem of libraries and tools, including TensorFlow Lite for mobile and embedded devices and TensorFlow Extended (TFX) for end-to-end ML pipelines.

    - **Example Use Case:** TensorFlow is used by researchers to develop new AI algorithms and by companies to deploy scalable AI solutions in production environments.

- **PyTorch:** PyTorch is an open-source machine learning library developed by Facebook. It is popular for its dynamic computational graph and ease of use, making it a favorite among researchers and developers.
    - **Features:**
        - **Dynamic Computational Graph:** PyTorch's dynamic graph allows for more flexibility and ease of debugging.
        - **Intuitive Interface:** PyTorch provides a simple and intuitive API, which makes it easy to learn and use.
        - **Integration with Python:** PyTorch integrates seamlessly with Python, making it a versatile tool for AI development.
    - **Example Use Case:** PyTorch is extensively used in academic research to develop and experiment with new neural network architectures and algorithms.
- **Keras:** Keras is an open-source neural network library that runs on top of other frameworks like TensorFlow and Theano. It is designed to enable fast experimentation and prototyping.
    - **Features:**
        - **User-Friendly API:** Keras offers a high-level, user-friendly API, which simplifies the process of building and training neural networks.

- **Modularity:** Keras allows users to create models by connecting building blocks like layers, optimizers, and activation functions.

- **Integration:** Keras can be used with multiple backends, providing flexibility in choosing the underlying computation engine.

- **Example Use Case:** Keras is often used by beginners and professionals for rapid prototyping of deep learning models due to its simplicity and ease of use.

- **MOTMSDD** The Metaverse of the Minds Social Direct Democracy (MOTMSDD) approach leverages AI software frameworks to revolutionize public engagement in policymaking. By integrating advanced AI algorithms, brain-computer interfaces (BCIs), and blockchain technology, MOTMSDD creates a digital metaverse where individuals are represented by digital twins. These digital twins participate in real-time, direct decision-making processes, ensuring that public policies reflect the true needs and preferences of the population. AI software frameworks analyse vast amounts of data from these interactions, providing insights into public sentiment and facilitating the resolution of conflicts between stakeholders. This approach not only enhances the inclusivity and transparency of public engagement but also ensures that policy decisions are data-driven and optimized for the holistic welfare of the community. Through MOTMSDD, citizens are empowered to actively participate in the governance process, fostering a more democratic and responsive society.

Other Notable Frameworks:

- **Scikit-learn:** A widely used machine learning library in Python, providing simple and efficient tools for data mining and analysis.
- **MXNet:** An open-source deep learning framework designed for efficiency and flexibility, used by Amazon for its deep learning services.
- **Caffe:** A deep learning framework focused on speed and modularity, commonly used for image classification tasks.

Summary: The convergence of powerful hardware and sophisticated software frameworks has driven the rapid advancements in AI. GPUs and TPUs provide the computational muscle needed to train complex models, while frameworks like TensorFlow, PyTorch, and Keras offer the tools and interfaces necessary for developing and deploying AI applications. Understanding the infrastructure that supports AI is crucial for leveraging its full potential and driving innovation across various industries.

This detailed exploration of AI infrastructure highlights the critical role of specialized hardware and software frameworks in the development and deployment of AI models. By understanding the hardware requirements and popular AI frameworks, we can appreciate the infrastructure that supports AI and enables its transformative potential in reshaping industries and everyday life.

2.3 Current Advancements in AI

2.3.1 2.3.1 AI in Industry

Artificial intelligence (AI) has become a transformative force across various industries, driving innovation and efficiency. This section explores the significant impact of AI in three key sectors: healthcare, finance, and automotive. Through detailed explanations and case studies, we illustrate how AI is revolutionizing these industries and shaping their future.

Healthcare

AI Applications in Medical Diagnosis: AI has made significant strides in medical diagnosis, assisting healthcare professionals in identifying diseases and conditions with high accuracy.

- **Imaging and Radiology:** AI algorithms, particularly those based on deep learning, are used to analyse medical images such as X-rays, MRIs, and CT scans. AI can detect abnormalities and diseases like cancer, fractures, and neurological disorders, often with greater precision than human radiologists.
- **Pathology:** AI systems can examine tissue samples and cellular images to diagnose diseases. For example, AI can identify cancerous cells in histopathology slides, aiding pathologists in making accurate diagnoses.

Example: IBM Watson for Oncology IBM Watson for Oncology uses AI to analyse patient data and medical literature to provide evidence-based treatment recommendations. It assists oncologists in developing personalized treatment plans for cancer patients.

Personalized Medicine: AI enables the development of personalized treatment plans tailored to individual patients' genetic profiles, medical histories, and lifestyle factors.

- **Genomics:** AI algorithms analyse genomic data to identify genetic mutations and their implications for disease susceptibility and treatment responses.

- **Predictive Analytics:** AI models predict disease progression and treatment outcomes based on patient data, allowing for more targeted and effective interventions.

Example: DeepMind's AlphaFold DeepMind's AlphaFold uses AI to predict protein structures with high accuracy. This breakthrough has significant implications for drug discovery and personalized medicine, as understanding protein structures is crucial for developing targeted therapies.

Healthcare Management: AI streamlines healthcare management by optimizing administrative processes, managing patient records, and improving operational efficiency.

- **Electronic Health Records (EHR):** AI systems automate the extraction and analysis of data from EHRs, reducing administrative burdens and improving data accuracy.
- **Resource Allocation:** AI optimizes the allocation of resources such as staff, equipment, and hospital beds, ensuring efficient operations and better patient care.

Example: Mayo Clinic's Clinical Decision Support System Mayo Clinic uses an AI-powered clinical decision support system to analyse patient data and provide real-time recommendations for care. This system improves diagnostic accuracy and enhances patient outcomes.

Finance

AI in Fraud Detection: AI plays a crucial role in identifying and preventing fraudulent activities in the financial sector.

- **Pattern Recognition:** AI algorithms analyse transaction patterns to detect anomalies that may indicate fraudulent behavior. These systems continuously learn from new data, improving their accuracy over time.

- **Real-Time Alerts:** AI systems provide real-time alerts for suspicious activities, allowing financial institutions to take immediate action to prevent fraud.

Example: PayPal's Fraud Detection System PayPal uses AI to monitor transactions and detect fraudulent activities. Its AI system analyses millions of transactions in real-time, identifying patterns that suggest fraud and blocking suspicious transactions.

Trading Algorithms: AI-driven trading algorithms, also known as algorithmic trading or high-frequency trading, execute trades at high speeds based on market data analysis.

- **Market Analysis:** AI algorithms analyse vast amounts of market data, including stock prices, trading volumes, and economic indicators, to identify trading opportunities.
- **Automated Execution:** AI systems execute trades automatically based on predefined criteria, optimizing timing and minimizing human error.

Example: Renaissance Technologies Renaissance Technologies, a hedge fund, uses AI and quantitative models to manage its investment strategies. Its Medallion Fund has achieved remarkable returns, largely attributed to its sophisticated AI-driven trading algorithms.

Risk Management: AI enhances risk management by identifying potential risks and providing insights for mitigating them.

- **Credit Scoring:** AI models assess credit risk by analysing borrowers' financial histories, spending patterns, and other relevant data.

- **Portfolio Management:** AI systems evaluate investment portfolios to identify risks and recommend adjustments to optimize returns and minimize losses.

Example: ZestFinance ZestFinance uses AI to improve credit scoring and underwriting processes. Its AI models analyse alternative data sources to assess creditworthiness, providing more accurate and inclusive credit assessments.

Automotive

Development of Autonomous Vehicles: AI is at the core of developing autonomous vehicles, enabling them to navigate and operate without human intervention.

- **Perception:** AI systems process data from sensors such as cameras, lidar, and radar to perceive the vehicle's surroundings, including detecting objects, pedestrians, and other vehicles.
- **Decision Making:** AI algorithms make real-time decisions based on sensor data, determining the vehicle's actions such as steering, acceleration, and braking.
- **Control Systems:** AI controls the vehicle's movements, ensuring safe and efficient operation.

Example: Tesla's Autopilot Tesla's Autopilot uses AI to enable semi-autonomous driving. The system processes data from multiple sensors and cameras to assist with steering, acceleration, and braking, enhancing safety and convenience for drivers.

Smart Transportation Systems: AI contributes to the development of smart transportation systems that optimize traffic flow and improve urban mobility.

- **Traffic Management:** AI analyses traffic patterns and predicts congestion, enabling dynamic traffic signal adjustments to improve flow.

- **Public Transportation:** AI optimizes public transportation routes and schedules, enhancing efficiency and reducing wait times for passengers.

Example: Singapore's Smart Traffic System Singapore uses an AI-powered smart traffic system to manage congestion and improve traffic flow. The system analyses real-time traffic data and adjusts traffic signals dynamically to reduce delays and enhance mobility.

Conclusion: AI is revolutionizing various industries by enhancing capabilities, improving efficiency, and enabling new possibilities. In healthcare, finance, and automotive sectors, AI applications are driving significant advancements, from medical diagnostics and personalized medicine to fraud detection, autonomous vehicles, and smart transportation systems. As AI continues to evolve, its impact will expand, reshaping industries and everyday life in profound ways.

This detailed exploration of AI in industry provides a comprehensive understanding of how AI is transforming key sectors such as healthcare, finance, and automotive. By examining specific applications and case studies, we can appreciate the diverse and far-reaching impact of AI on our world.

2.3.2 Breakthroughs in AI Research

AI research has seen remarkable advancements in recent years, resulting in groundbreaking achievements that have pushed the boundaries of what is possible with artificial intelligence. This section explores significant milestones such as AlphaGo, AlphaFold, and GPT-4, and discusses the ongoing efforts to address ethical concerns, bias, and fairness in AI systems.

AlphaGo and Beyond

AlphaGo: AlphaGo, developed by DeepMind, is a landmark achievement in AI research. In 2016, AlphaGo became the first AI system to defeat a world champion Go player, Lee Sedol, in a five-game match. This victory was significant due to the complexity of Go, which has more possible moves than there are atoms in the universe.

- **How AlphaGo Works:**
    - **Reinforcement Learning:** AlphaGo uses reinforcement learning, where the system learns optimal strategies through self-play, receiving rewards for successful moves.
    - **Deep Neural Networks:** It employs deep neural networks to evaluate board positions and predict the most promising moves.
    - **Monte Carlo Tree Search (MCTS):** AlphaGo combines neural networks with MCTS to simulate potential future moves and select the best actions.

Impact of AlphaGo: AlphaGo's success demonstrated the power of combining deep learning with reinforcement learning and has inspired further research in AI and game theory. It also highlighted AI's potential to tackle complex, strategic problems beyond traditional computational capabilities.

AlphaZero: Building on AlphaGo, DeepMind developed AlphaZero, a more generalized version that can master multiple games like chess, shogi, and Go without prior knowledge, learning solely through self-play. AlphaZero's versatility marks a significant step towards more adaptable AI systems.

AlphaFold: Another groundbreaking achievement by DeepMind is AlphaFold, an AI system designed to predict protein structures. In 2020, AlphaFold demonstrated its ability to determine the 3D shapes of proteins with remarkable accuracy, a problem that has challenged scientists for decades.

- **How AlphaFold Works:**
    - **Deep Learning Models:** AlphaFold uses deep learning models to predict the distances between amino acids and the angles of chemical bonds within proteins.
    - **Training on Protein Data:** The system was trained on vast amounts of protein data, enabling it to learn the patterns and rules governing protein folding.

Impact of AlphaFold: AlphaFold's ability to accurately predict protein structures has profound implications for biology and medicine. It accelerates the understanding of diseases, the development of new drugs, and the creation of synthetic proteins for various applications.

GPT-4: GPT-4, developed by OpenAI, is one of the most advanced language models to date. It builds on the success of its predecessors and offers significant improvements in language understanding, generation, and contextual comprehension.

- **How GPT-4 Works:**
    - **Transformer Architecture:** GPT-4 uses an advanced version of the transformer architecture, which excels in processing sequential data and capturing long-range dependencies.
    - **Pre-training on Vast Data:** The model was pre-trained on a diverse corpus of internet text, allowing it to learn a wide range of language patterns and knowledge.

Impact of GPT-4: GPT-4's ability to generate coherent and contextually relevant text has numerous applications, from content creation to customer service automation. It also raises questions about the potential misuse of AI-generated text and the need for ethical guidelines.

AI Ethics and Fairness

Progress in Addressing Ethical Concerns: As AI systems become more integrated into society, addressing ethical concerns has become a critical area of research. Efforts are being made to ensure that AI technologies are developed and deployed responsibly, with a focus on transparency, accountability, and human rights.

- **Ethical AI Guidelines:** Organizations and governments are developing ethical guidelines for AI development. These guidelines emphasize the importance of fairness, transparency, and accountability in AI systems.
- **Bias Mitigation:** Researchers are working on methods to identify and mitigate biases in AI models. This includes developing techniques for bias detection, data augmentation, and algorithmic fairness.

Example: Fairness Indicators Google has developed Fairness Indicators, a toolkit for evaluating the fairness of machine learning models. It provides metrics and visualizations to assess model performance across different demographic groups, helping developers identify and address biases.

Addressing Bias in AI Systems: Bias in AI systems can arise from various sources, including biased training data, algorithmic design, and deployment contexts. Addressing these biases is essential to ensure that AI systems are fair and equitable.

- **Data Bias:** Ensuring that training data is representative and free from biases is crucial. This involves collecting diverse datasets and preprocessing data to remove biases.
- **Algorithmic Fairness:** Developing algorithms that are inherently fair and can adjust for biases in data. Techniques such as adversarial debiasing and fairness constraints are used to create more equitable models.

- **Transparency and Accountability:** Implementing measures to make AI systems transparent and accountable. This includes documenting the decision-making processes of AI models and providing explanations for their outputs.

Example: IBM AI Fairness 360 IBM AI Fairness 360 is an open-source toolkit that provides metrics and algorithms to help researchers and developers detect and mitigate bias in AI models. It includes tools for assessing fairness at various stages of the AI lifecycle.

Fairness in AI Deployment: Ensuring fairness in the deployment of AI systems involves continuous monitoring and evaluation to identify and address any emerging biases or ethical concerns.

- **Monitoring and Evaluation:** Regularly monitoring AI systems in real-world environments to detect biases and ensure fair outcomes.
- **Stakeholder Engagement:** Engaging with diverse stakeholders, including marginalized communities, to understand the impacts of AI systems and ensure that their perspectives are considered in AI development and deployment.

Example: Inclusive AI Development Organizations are adopting inclusive AI development practices that involve collaboration with diverse communities to ensure that AI systems are designed and deployed in ways that are fair and beneficial for all.

This detailed exploration of breakthroughs in AI research highlights significant achievements such as AlphaGo, AlphaFold, and GPT-4, and emphasizes the importance of addressing ethical concerns, bias, and fairness in AI systems. By understanding these advancements and challenges, we can appreciate the transformative potential of AI and work towards ensuring that it benefits all members of society.

2.3.3 AI in Everyday Life

Artificial Intelligence (AI) is increasingly becoming a part of our daily lives, offering convenience, enhancing experiences, and transforming how we interact with technology. This section explores how AI is integrated into consumer applications and education, highlighting its impact on personal assistants, recommendation systems, smart home devices, and personalized learning platforms.

Consumer Applications

Personal Assistants: AI-powered personal assistants like Siri, Alexa, and Google Assistant have revolutionized how we interact with our devices. These assistants use natural language processing (NLP) and machine learning to understand and respond to user queries, perform tasks, and provide information.

- **Capabilities:**
    - **Voice Recognition:** AI assistants can recognize and interpret spoken commands, making it easy for users to interact with their devices hands-free.
    - **Task Automation:** They can perform a variety of tasks such as setting reminders, sending messages, making calls, and controlling smart home devices.
    - **Information Retrieval:** Personal assistants can answer questions by retrieving information from

the web, providing weather updates, news, and other relevant data.

Example: Amazon Alexa Amazon Alexa is integrated into numerous devices, allowing users to control smart home gadgets, play music, order products online, and more. Alexa's capabilities are continually expanding through third-party "skills" developed by external developers.

Recommendation Systems: AI-driven recommendation systems are prevalent in e-commerce, streaming services, and social media platforms. These systems analyse user behavior and preferences to suggest products, movies, music, and content that users are likely to enjoy.

- **Techniques Used:**
    - **Collaborative Filtering:** This technique makes recommendations based on the preferences of similar users.
    - **Content-Based Filtering:** It recommends items that are similar to those a user has liked in the past.
    - **Hybrid Approaches:** Combining collaborative and content-based filtering to improve recommendation accuracy.

Example: Netflix Recommendation System Netflix uses a sophisticated recommendation system that analyses viewing history, user ratings, and preferences to suggest movies and TV shows. This personalized approach enhances user engagement and satisfaction.

Smart Home Devices: AI is integral to the functionality of smart home devices, which automate and enhance the living experience. These devices include smart thermostats, lights, security cameras, and appliances that can be controlled remotely or through voice commands.

- **Capabilities:**
    - **Automation:** AI enables devices to learn user preferences and routines, automating tasks such as adjusting temperature, lighting, and security settings.
    - **Energy Efficiency:** Smart thermostats like the Nest Learning Thermostat use AI to optimize heating and cooling schedules, reducing energy consumption and costs.
    - **Security:** AI-powered security cameras and systems can detect unusual activities and send alerts, enhancing home security.

Example: Google Nest Google Nest offers a range of smart home products, including thermostats, cameras, and doorbells. These devices use AI to provide personalized, energy-efficient, and secure home automation solutions.

Education and Learning

AI-Driven Personalized Learning Platforms: AI is transforming education by providing personalized learning experiences tailored to individual students' needs and learning styles. These platforms use data analytics and machine learning to adapt content, pace, and difficulty based on student performance.

- **Capabilities:**
    - **Adaptive Learning:** AI systems adjust the learning path in real-time, providing additional resources or challenges as needed.
    - **Feedback and Assessment:** They offer instant feedback and assessments, helping students understand their strengths and areas for improvement.
    - **Engagement:** Interactive and engaging content such as gamified lessons and virtual tutors keep students motivated and interested in learning.

Example: Khan Academy Khan Academy uses AI to personalize the learning experience for students. The platform provides tailored exercises and instructional videos, tracking progress and suggesting resources to help students master concepts.

Educational Tools: AI-powered tools are enhancing the educational experience for both students and educators. These tools include intelligent tutoring systems, automated grading, and language translation services.

- **Intelligent Tutoring Systems:** These systems provide one-on-one tutoring, using AI to mimic the guidance of a human tutor. They can answer questions, provide explanations, and adapt to the student's learning pace.
    - **Example: Carnegie Learning** Carnegie Learning offers AI-driven math tutoring software that provides personalized instruction and practice.

The system adapts to each student's learning style and needs, offering targeted support.

- **Automated Grading:** AI systems can grade assignments and exams quickly and accurately, providing instant feedback to students and reducing the administrative burden on educators.

    - **Example: Gradescope** Gradescope uses AI to assist with grading, enabling educators to grade assignments more efficiently. The platform supports various types of assessments, including essays and coding assignments.

- **Language Translation Services:** AI-powered translation tools help students and educators overcome language barriers, facilitating communication and access to educational resources in different languages.

    - **Example: Google Translate** Google Translate uses neural machine translation to provide accurate and contextually appropriate translations, supporting over 100 languages. It is widely used in educational settings to help students understand foreign language materials.

Case Study: Duolingo Duolingo is a language learning platform that uses AI to personalize lessons and track progress. The app adapts to the learner's proficiency level, providing targeted practice and feedback to help users achieve language fluency.

This detailed exploration of AI in everyday life highlights how AI is integrated into consumer applications and education, enhancing convenience, personalization, and efficiency. By examining specific examples and use cases, we can appreciate the transformative potential of AI in shaping our daily experiences and improving the way we learn and interact with technology.

2.4 Future Potential of AI

2.4.1 AI in Scientific Discovery

Artificial Intelligence (AI) is revolutionizing scientific research by accelerating the pace of discovery and opening new frontiers across various disciplines. This section explores how AI is transforming scientific research and its potential to solve complex problems in physics, chemistry, and biology.

Accelerating Research

How AI is Transforming Scientific Research: AI technologies, particularly machine learning and deep learning, are enhancing scientific research by automating data analysis, uncovering patterns, and generating hypotheses. These capabilities significantly accelerate the research process, allowing scientists to focus on higher-level problem-solving and innovation.

- **Data Analysis and Interpretation:**
    - **Automated Analysis:** AI algorithms can process vast amounts of data quickly and accurately, identifying patterns and correlations that might be missed by human researchers.
    - **Predictive Modeling:** Machine learning models can predict outcomes based on existing data, helping researchers design better experiments and anticipate results.

Example: AI in Genomics AI is playing a crucial role in genomics by analysing genetic data to identify disease-related genes and understand genetic variations. For instance, AI-driven tools can analyse whole-genome sequencing data to predict the likelihood of genetic disorders, enabling early diagnosis and personalized treatment.

- **High-Throughput Screening:**
    - **Drug Discovery:** AI accelerates drug discovery by analysing chemical compounds and predicting their interactions with biological targets. This reduces the time and cost associated with traditional experimental methods.
    - **Molecular Simulations:** AI can simulate molecular interactions at a scale and speed unattainable by conventional methods, providing insights into drug efficacy and potential side effects.

Example: Atomwise Atomwise uses AI to predict the binding affinity of small molecules to target proteins, aiding in the discovery of new drugs. Their AI-driven approach has led to the identification of promising compounds for diseases such as Ebola and multiple sclerosis.

New Frontiers: Potential for AI to Solve Complex Problems

Physics: AI is pushing the boundaries of physics by helping scientists analyse data from experiments, simulations, and observations, leading to new discoveries and deeper understanding.

- **Quantum Mechanics:** AI algorithms are used to solve complex quantum mechanical problems, such as optimizing quantum circuits and simulating quantum systems. This contributes to advancements in quantum computing and our understanding of fundamental physical phenomena.
- **Astrophysics:** AI assists in analysing astronomical data, identifying exoplanets, and detecting gravitational waves.

Machine learning models can process data from telescopes and space missions, uncovering insights about the universe.

Example: AI in Gravitational Wave Detection AI has been instrumental in analysing data from the Laser Interferometer Gravitational-Wave Observatory (LIGO). Machine learning algorithms help identify gravitational wave signals from the noise, enabling the detection of events such as black hole mergers.

Chemistry: AI is transforming chemistry by enabling the discovery of new materials and the understanding of chemical processes at an unprecedented level.

- **Materials Science:** AI accelerates the discovery of new materials with desired properties, such as high-strength alloys, superconductors, and efficient catalysts. Machine learning models predict the behavior of materials under different conditions, guiding experimental efforts.
- **Chemical Synthesis:** AI helps design efficient synthetic routes for chemical compounds, optimizing reaction conditions and reducing the need for trial-and-error experimentation.

Example: AI in Material Discovery DeepMind's AI has been used to predict the stability of crystal structures, aiding in the discovery of new materials. This approach has the potential to revolutionize industries ranging from electronics to energy storage.

Biology: AI is making significant contributions to biology by providing insights into complex biological systems and processes.

- **Proteomics:** AI tools like AlphaFold predict protein structures with high accuracy, providing insights into their functions and interactions. This knowledge is crucial for understanding diseases and developing targeted therapies.
- **Ecology:** AI models analyse ecological data to monitor biodiversity, track wildlife populations, and predict the

impacts of climate change. This helps in the conservation of endangered species and ecosystems.

Example: AlphaFold DeepMind's AlphaFold has revolutionized structural biology by accurately predicting protein structures. This breakthrough has implications for drug discovery, understanding disease mechanisms, and designing novel proteins with specific functions.

Conclusion: AI is not only accelerating the pace of scientific research but also enabling breakthroughs that were previously thought to be beyond reach. By automating data analysis, generating predictive models, and uncovering hidden patterns, AI is transforming the landscape of scientific discovery in physics, chemistry, biology, and beyond. The integration of AI into scientific research promises to unlock new knowledge and address some of the most complex challenges facing humanity.

This comprehensive exploration of AI in scientific discovery highlights its transformative impact on research processes and its potential to solve complex problems across various scientific disciplines. By examining specific examples and use cases, we can appreciate how AI is driving innovation and advancing our understanding of the natural world.

2.4.2 AI and Creativity

Artificial Intelligence (AI) is not only revolutionizing scientific discovery and industry but also making significant strides in the realm of creativity. This section explores how AI is being used in art, music, and literature, and how it is augmenting human creativity and fostering collaboration.

Creative AI

Exploration of AI in Art, Music, and Literature: AI has become an innovative tool in the creative process, generating new forms of art, composing music, and even writing literature. These advancements showcase the potential of AI to create original works and push the boundaries of human creativity.

- **Art:** AI is capable of producing unique artworks by learning from existing styles and creating new ones. Generative Adversarial Networks (GANs) are often used to create realistic images and artworks that mimic different artistic styles.

    - **Example: DeepArt** DeepArt uses neural networks to transform photos into artworks in the style of famous artists like Van Gogh or Picasso. The AI analyses the artistic style and applies it to new images, creating visually stunning results.

    - **Example: AI Art Auction** In 2018, an AI-generated artwork titled "Portrait of Edmond de Belamy" was sold at Christie's auction for $432,500. The artwork was created using a GAN trained on a dataset of historical portraits.

- **Music:** AI can compose music by learning from a vast database of musical pieces, identifying patterns, and generating original compositions. This application of AI ranges from classical music to contemporary genres.

    - **Example: OpenAI's MuseNet** MuseNet is an AI model developed by OpenAI that can generate music in various styles and genres. It can create compositions that blend the styles of different artists, demonstrating the versatility of AI in music creation.

  - **Example: Amper Music** Amper Music is an AI-driven music composition platform that allows users to create custom music tracks for their projects. It offers tools for generating music based on user preferences and editing the compositions.

- **Literature:** AI is also making strides in writing, from generating poetry to drafting entire novels. Natural Language Processing (NLP) models, such as GPT-4, can produce coherent and contextually relevant text based on prompts.

  - **Example: AI-Generated Novels** In Japan, a novel partially written by an AI made it through the initial screening of a national literary competition. The AI wrote the narrative based on the structure and style provided by human authors, showcasing the potential for collaborative writing.

  - **Example: Poetry by AI** Platforms like PoemPortraits use AI to generate poetry based on user input. The AI analyses user-provided words and crafts poems that are often surprisingly creative and evocative.

Collaborative Creativity

The Role of AI in Augmenting Human Creativity and Collaboration: AI is not just a standalone creator; it also plays a crucial role in augmenting human creativity and fostering collaboration between humans and machines. By working together, AI and humans can produce innovative and original works that neither could create alone.

- **Enhancing Creative Processes:** AI tools can assist artists, musicians, and writers in their creative processes by providing inspiration, suggesting new ideas, and automating repetitive tasks. This allows creators to focus more on conceptual and expressive aspects of their work.
  - **Example: Adobe Sensei** Adobe Sensei integrates AI into Adobe's suite of creative tools, offering features like automatic tagging of photos, content-aware fill, and style transfer. These tools enhance the productivity and creativity of designers and photographers.
  - **Example: Jukedeck** Jukedeck uses AI to help musicians compose original music by generating base tracks that artists can build upon. This collaboration between AI and musicians leads to new and innovative musical pieces.
- **Collaborative Art Projects:** AI enables new forms of collaborative art projects where human creativity and machine intelligence come together to produce unique works. These projects often explore the interplay between human intuition and algorithmic generation.
  - **Example: The Next Rembrandt** The Next Rembrandt is a project that used AI to create a new painting in the style of Rembrandt. The AI

analysed Rembrandt's works to generate a new, original piece that mimicked his style. This project showcased the potential of AI to collaborate with human artists in preserving and extending artistic legacies.

- **Example: AI and Human Co-Creation in Film** AI is being used in the film industry to assist in scriptwriting, editing, and even directing. For example, the short film "Zone Out" was co-written by an AI, demonstrating how AI can contribute to the storytelling process and collaborate with filmmakers.

- **Education and Training:** AI can also be used to educate and train artists, musicians, and writers by providing personalized feedback and suggestions. AI-powered platforms offer interactive learning experiences that adapt to the user's skill level and progress.

  - **Example: Smart Sparrow** Smart Sparrow is an adaptive learning platform that uses AI to provide personalized feedback to students in creative fields. It helps learners improve their skills by offering targeted exercises and critiques.

  - **Example: AIVA (Artificial Intelligence Virtual Artist)** AIVA is an AI that composes classical music and provides insights into music theory and composition. It serves as both a creative collaborator and an educational tool for musicians.

Conclusion: AI is revolutionizing the creative industries by generating original works of art, music, and literature, and by augmenting human creativity through collaboration. As AI continues to evolve, its role in the creative process will expand, leading to new forms of artistic expression and innovation. By working together, humans and AI can push the boundaries of creativity and produce works that reflect the unique contributions of both.

This comprehensive exploration of AI in creativity highlights its transformative impact on art, music, and literature, and emphasizes the collaborative potential between AI and human creators. By examining specific examples and use cases, we can appreciate how AI is enhancing and expanding the creative landscape.

2.4.3 AI in Public Policy and Governance

Artificial Intelligence (AI) is poised to transform public policy and governance, offering new tools for urban planning, traffic management, public safety, and data-driven policy analysis. This section explores how AI is being applied in these areas to create smarter cities and enhance decision-making processes in governance.

Smart Cities

AI Applications in Urban Planning: AI is revolutionizing urban planning by enabling data-driven approaches to designing and managing urban spaces. Through the analysis of vast amounts of data, AI helps urban planners make informed decisions that improve the quality of life for residents.

- **Predictive Modeling:** AI can predict urban growth and development patterns by analysing historical data and current trends. This helps planners design cities that are more sustainable and resilient.
- **Resource Allocation:** AI optimizes the allocation of resources such as water, energy, and public services by analysing consumption patterns and predicting future needs.
- **Environmental Impact Assessment:** AI models assess the environmental impact of urban development projects, helping planners minimize negative effects on ecosystems.

Example: Sidewalk Labs Sidewalk Labs, a subsidiary of Alphabet Inc., uses AI to design and implement smart city solutions. Their projects include optimizing urban mobility, improving energy efficiency, and creating sustainable urban environments.

Traffic Management: AI enhances traffic management systems by analysing real-time traffic data, predicting congestion, and optimizing traffic flow. This reduces travel time, lowers emissions, and improves overall transportation efficiency.

- **Real-Time Traffic Monitoring:** AI systems monitor traffic conditions in real-time using data from sensors, cameras, and GPS devices. This information is used to adjust traffic signals and manage traffic flow dynamically.
- **Predictive Analytics:** AI models predict traffic congestion and identify potential bottlenecks before they occur. This allows for proactive measures to be taken, such as adjusting traffic signal timings or rerouting traffic.
- **Autonomous Vehicles:** AI-powered autonomous vehicles are integrated into traffic management systems to improve safety and efficiency. These vehicles can communicate with each other and with traffic infrastructure to optimize routes and reduce congestion.

Example: Singapore's Intelligent Transport System Singapore's Intelligent Transport System uses AI to manage traffic flow, monitor road conditions, and provide real-time traffic updates to commuters.

This system has significantly reduced congestion and improved transportation efficiency in the city-state.

Public Safety: AI applications in public safety include crime prediction, emergency response management, and disaster preparedness. By analysing data from various sources, AI helps authorities make informed decisions to enhance public safety.

- **Crime Prediction and Prevention:** AI analyses crime data to identify patterns and predict future criminal activity. This information is used to deploy law enforcement resources more effectively and prevent crime.
- **Emergency Response:** AI systems optimize emergency response by analysing data from 911 calls, social media, and sensors. This helps dispatchers prioritize incidents and allocate resources efficiently.
- **Disaster Preparedness:** AI models predict natural disasters such as earthquakes, floods, and hurricanes by analysing geological and meteorological data. This allows for timely warnings and better preparedness.

Example: PredPol PredPol, short for Predictive Policing, uses AI to analyse crime data and predict where crimes are likely to occur. Law enforcement agencies use these predictions to focus patrols in high-risk areas, helping to prevent crime before it happens.

Policy Making

Use of AI for Data-Driven Policy Analysis and Decision-Making: AI is transforming policy making by providing data-driven insights that inform decisions and improve outcomes. By analysing large datasets, AI helps policymakers understand complex issues, evaluate policy impacts, and design effective interventions.

- **Policy Analysis:** AI models analyse the potential impacts of policy proposals by simulating different scenarios and outcomes. This helps policymakers understand the trade-offs and make informed decisions.

- **Sentiment Analysis:** AI analyses public opinion data from social media, surveys, and other sources to gauge public sentiment on policy issues. This information helps policymakers understand public concerns and preferences.
- **Economic Forecasting:** AI models predict economic trends and assess the potential impacts of economic policies. This helps governments design policies that promote growth, stability, and social welfare.

Example: AI for Healthcare Policy AI is used to analyse healthcare data and inform policy decisions aimed at improving public health. For instance, AI models can predict the spread of infectious diseases, evaluate the effectiveness of interventions, and optimize healthcare resource allocation.

Case Study: AI in Environmental Policy AI is used to analyse environmental data and inform policies aimed at combating climate change. For example, AI models predict the impact of carbon pricing on emissions, helping policymakers design effective climate policies.

- **Environmental Monitoring:** AI systems monitor environmental conditions in real-time, providing data on air and water quality, deforestation, and wildlife populations. This information helps policymakers design policies that protect the environment.
- **Climate Modeling:** AI enhances climate models by analysing vast amounts of meteorological data. This helps predict the impacts of climate change and design policies that mitigate its effects.

Example: The AI for Earth Program Microsoft's AI for Earth program uses AI to address global environmental challenges. The program supports projects that use AI to monitor and manage natural resources, track wildlife populations, and predict climate change impacts.

Conclusion: AI is transforming public policy and governance by enabling data-driven approaches to urban planning, traffic management, public safety, and policy analysis. By leveraging AI, governments can make more informed decisions, design effective policies, and improve the quality of life for their citizens. As AI technology continues to evolve, its impact on public policy and governance will only grow, leading to smarter, more efficient, and more responsive governments.

This comprehensive exploration of AI in public policy and governance highlights its transformative impact on urban planning, traffic management, public safety, and policy analysis. By examining specific examples and case studies, we can appreciate how AI is enhancing decision-making processes and creating smarter cities and more effective governance.

2.5 Ethical and Societal Implications of AI

2.5.1 Ethical Challenges

As artificial intelligence (AI) becomes increasingly integrated into various aspects of society, addressing its ethical challenges is paramount. Ensuring that AI systems are fair, transparent, and accountable is crucial to their responsible development and deployment. This section delves into the ethical challenges of bias and discrimination in AI models and the importance of transparency and accountability in AI systems.

Bias and Discrimination

Addressing Bias in AI Models: Bias in AI models can arise from various sources, including biased training data, algorithmic design, and the deployment context. These biases can lead to unfair and discriminatory outcomes, which are particularly problematic in critical areas such as hiring, lending, law enforcement, and healthcare.

- **Sources of Bias:**
  - **Training Data:** AI models learn from data, and if this data contains historical biases or is not representative of the diverse population, the model will likely perpetuate these biases.
  - **Algorithm Design:** The way algorithms are designed can introduce biases. For instance, certain features may be weighted more heavily than others, leading to biased outcomes.
  - **Deployment Context:** The environment in which an AI system is deployed can also contribute to bias. Societal and institutional biases can influence the performance and fairness of AI systems.

Strategies to Ensure Fairness:

- **Diverse and Representative Data:** Ensuring that training datasets are diverse and representative of all relevant groups is crucial. This involves collecting data from various demographics and ensuring that minority groups are adequately represented.
- **Bias Detection and Mitigation:** Implementing techniques to detect and mitigate bias in AI models is essential. This includes using fairness metrics to evaluate model performance across different groups and applying algorithms designed to reduce bias.
- **Continuous Monitoring:** AI systems should be continuously monitored for bias throughout their lifecycle. This involves regularly auditing the models and updating them as new data becomes available.

Example: Fairness Indicators Google's Fairness Indicators is a tool that helps developers assess and mitigate bias in their machine learning models. It provides metrics and visualizations to evaluate model performance across different demographic groups, ensuring that AI systems are fair and equitable.

Transparency and Accountability

Importance of Explainable AI: Explainable AI (XAI) refers to methods and techniques that make the decision-making processes of AI systems understandable to humans. Ensuring that AI decisions are transparent and explainable is critical for building trust and accountability.

- **Challenges of Black-Box Models:**
    - Many AI models, particularly deep learning models, are often described as "black boxes" because their internal workings are not easily interpretable. This lack of transparency can lead to challenges in understanding how decisions are made and identifying potential biases.
    - Black-box models can make it difficult to explain why certain decisions were made, which is especially problematic in high-stakes areas such as healthcare, finance, and criminal justice.

Strategies for Explainable AI:

- **Interpretable Models:** Developing models that are inherently interpretable, such as decision trees or linear models, can enhance transparency. While these models may not be as powerful as complex neural networks, they provide insights into the decision-making process.
- **Post-Hoc Explanations:** Techniques such as LIME (Local Interpretable Model-Agnostic Explanations) and SHAP (SHapley Additive exPlanations) provide post-hoc explanations for black-box models. These methods help interpret the predictions of complex models by approximating them with simpler, interpretable models.

- **Visualization Tools:** Visualizing the internal workings and decisions of AI models can help stakeholders understand how the models function. Tools that map feature importance and decision paths can enhance the interpretability of AI systems.

Example: IBM's AI Explainability 360 IBM's AI Explainability 360 is an open-source toolkit that provides a range of algorithms and metrics to help developers create explainable AI models. It includes tools for generating post-hoc explanations, evaluating the interpretability of models, and improving transparency.

Accountable Decision-Making: Ensuring accountability in AI systems involves establishing clear responsibilities and processes for overseeing AI development and deployment. This is crucial for addressing the ethical implications of AI and maintaining public trust.

- **Accountability Mechanisms:**
    - **Governance Structures:** Establishing governance structures that oversee AI development and deployment can ensure that ethical standards are maintained. This includes creating ethics boards and appointing AI ethics officers.
    - **Regulatory Compliance:** AI systems should comply with relevant regulations and standards. Regulatory bodies play a key role in setting guidelines for the ethical use of AI and ensuring that organizations adhere to these standards.
    - **Audit and Oversight:** Regular audits of AI systems can help identify and address ethical issues. Independent oversight bodies can provide impartial assessments of AI systems and their impacts.

Example: GDPR and AI Accountability The General Data Protection Regulation (GDPR) in the European Union includes provisions for AI accountability. It requires organizations to provide explanations for automated decisions and ensure that individuals have the right to contest and seek human intervention in AI-driven decisions.

Conclusion: Addressing the ethical challenges of bias, discrimination, transparency, and accountability in AI is essential for the responsible development and deployment of AI systems. By implementing strategies to ensure fairness, developing explainable AI, and establishing accountable decision-making processes, we can build AI systems that are trustworthy, equitable, and beneficial for all members of society.

This comprehensive exploration of ethical challenges in AI highlights the importance of addressing bias, ensuring transparency, and maintaining accountability. By examining specific examples and strategies, we can understand how to develop and deploy AI systems that are ethically sound and socially responsible.

2.5.2 Impact on Employment

The rise of artificial intelligence (AI) is poised to transform the employment landscape, bringing both challenges and opportunities. This section explores the potential for AI to automate jobs and the implications for the workforce, as well as the emerging job roles and opportunities created by AI.

Job Displacement

Potential for AI to Automate Jobs: AI technologies, particularly those involving machine learning, robotics, and natural language processing, have the potential to automate a wide range of tasks across various industries. This automation can lead to significant job displacement, particularly in roles involving repetitive and routine tasks.

- **Automation of Routine Tasks:**
    - **Manufacturing:** AI-powered robots and automation systems are increasingly used in manufacturing to perform tasks such as assembly, welding, and quality control. This reduces the need for human labor in repetitive and hazardous tasks.
    - **Customer Service:** AI chatbots and virtual assistants can handle customer inquiries, process orders, and provide support, reducing the need for human customer service representatives.
    - **Data Entry and Analysis:** AI algorithms can automate data entry, processing, and analysis, making roles that involve these tasks susceptible to automation.

Example: Amazon's Fulfillment Centers Amazon's fulfillment centers use AI-driven robots to sort and transport packages, significantly increasing efficiency and reducing the need for human labor in these roles. While this automation improves productivity, it also displaces workers who previously performed these tasks.

Implications for the Workforce: The automation of jobs by AI has several implications for the workforce, including potential unemployment, the need for retraining, and shifts in the types of skills that are in demand.

- **Unemployment Risk:** Workers in roles that are highly susceptible to automation face the risk of job loss. This is particularly concerning for low-skill jobs, where

workers may find it challenging to transition to new roles.
- **Skill Shifts:** As AI automates routine tasks, there will be a growing demand for skills that complement AI, such as problem-solving, critical thinking, and creativity. Workers will need to adapt by acquiring new skills that are less likely to be automated.
- **Retraining and Upskilling:** Governments, educational institutions, and companies will need to invest in retraining and upskilling programs to help workers transition to new roles. This involves providing access to education and training in emerging fields and technologies.

Example: The European Union's Digital Skills and Jobs Coalition The European Union has launched initiatives such as the Digital Skills and Jobs Coalition to address the skill gaps created by digital transformation. These programs aim to provide training in digital skills and help workers transition to new roles in the digital economy.

New Opportunities
Emerging Job Roles and Opportunities Created by AI: While AI is automating certain tasks, it is also creating new job roles and opportunities. These emerging roles often require specialized skills in AI development, implementation, and maintenance, as well as skills that leverage human creativity and emotional intelligence.

- **AI Development and Implementation:**
  - **Data Scientists and Analysts:** As organizations leverage AI, there is a growing demand for data

scientists and analysts who can develop, train, and maintain AI models.

- **AI Engineers and Developers:** AI engineers and developers design and implement AI systems and applications. These roles require expertise in machine learning, deep learning, and software development.

- **AI Ethics and Policy Specialists:** As the ethical implications of AI become more prominent, there is a need for specialists who can address ethical concerns, develop policies, and ensure that AI systems are used responsibly.

Example: AI Engineers at OpenAI OpenAI employs a team of AI engineers and researchers who develop cutting-edge AI technologies and applications. These roles require deep expertise in AI and machine learning, as well as a commitment to ethical AI development.

- **Human-AI Collaboration:**

  - **AI Trainers:** AI systems often require human input to learn and improve. AI trainers provide this input by labeling data, providing feedback, and helping AI systems understand complex tasks.

  - **Human-AI Interaction Designers:** These designers focus on creating intuitive and effective interactions between humans and AI systems, ensuring that AI applications are user-friendly and meet human needs.

- **AI-Augmented Roles:** Many existing roles are being augmented by AI, where AI tools enhance human capabilities. For example, doctors use AI to assist in diagnosis, lawyers use AI for legal research, and marketers use AI for customer insights and campaign optimization.

Example: Doctors Using AI in Diagnosis In healthcare, doctors are using AI tools like IBM Watson Health to assist in diagnosing diseases and developing treatment plans. These AI systems analyse medical data and provide insights that enhance the doctors' decision-making capabilities.

Economic and Social Impacts: The integration of AI into the workforce will have broader economic and social impacts, including changes in industry dynamics, productivity gains, and shifts in the labor market.

- **Productivity Gains:** AI has the potential to significantly increase productivity by automating routine tasks and enhancing human capabilities. This can lead to economic growth and higher standards of living.
- **Industry Transformation:** Industries that adopt AI technologies are likely to experience significant transformation, with new business models and operational efficiencies. This can lead to the emergence of new markets and opportunities.
- **Social Equity:** Ensuring that the benefits of AI are distributed equitably is crucial. Policymakers and businesses must work together to address potential disparities and ensure that all segments of society can benefit from AI advancements.

Example: AI in Financial Services In the financial services industry, AI is transforming operations by automating trading, enhancing fraud detection, and improving customer service. This transformation is creating new roles in AI development and cybersecurity while also increasing productivity and efficiency.

Conclusion: The impact of AI on employment is multifaceted, with both challenges and opportunities. While AI has the potential to automate jobs and displace workers, it also creates new roles and opportunities that require specialized skills. By addressing the ethical challenges of bias, transparency, and accountability, and by investing in retraining and upskilling programs, we can ensure that the workforce is prepared for the changes brought about by AI. Embracing AI's potential while mitigating its risks will be key to shaping a future where AI enhances human capabilities and contributes to economic and social progress.

This detailed exploration of the impact of AI on employment highlights the potential for job displacement and the emerging opportunities created by AI. By examining specific examples and strategies, we can understand how to navigate the challenges and leverage the benefits of AI in the workforce.

2.5.3 Privacy and Security

As artificial intelligence (AI) continues to advance, it brings both tremendous opportunities and significant challenges, particularly in the areas of privacy and security. This section explores how to balance AI innovation with privacy protection and address security vulnerabilities and the potential for misuse of AI.

Data Privacy

Balancing AI Innovation with Privacy Protection: The development of AI relies heavily on vast amounts of data, which often includes personal and sensitive information. Ensuring data privacy while fostering AI innovation is a critical challenge that requires careful consideration and robust strategies.

- **Data Collection and Usage:**
    - **Transparency:** Organizations must be transparent about the data they collect, how it is used, and the purposes for which it is processed. Clear communication helps build trust with users and ensures compliance with privacy regulations.
    - **Consent:** Obtaining explicit consent from users before collecting and using their data is essential. This involves informing users about the data collection process and giving them control over their personal information.
    - **Data Minimization:** Collecting only the data that is necessary for the specific purpose of the AI application helps minimize privacy risks. Data minimization reduces the potential for misuse and limits the exposure of personal information.

Example: General Data Protection Regulation (GDPR) The GDPR, implemented in the European Union, sets strict guidelines for data protection and privacy. It requires organizations to be transparent about data collection, obtain explicit consent, and ensure the security of personal data. Compliance with GDPR is mandatory for organizations operating within the EU or handling the data of EU citizens.

- **Privacy-Preserving Techniques:**

- **Anonymization:** Removing personally identifiable information from datasets to protect individual privacy. Anonymized data can be used for AI training without exposing personal details.
- **Federated Learning:** A decentralized approach to training AI models that keeps data localized on users' devices while only sharing model updates. This method enhances privacy by ensuring that raw data never leaves the user's device.
- **Differential Privacy:** Adding noise to data in a controlled manner to protect individual privacy while still allowing useful insights to be derived from the dataset. Differential privacy provides mathematical guarantees that individual data points cannot be re-identified.

Example: Google's Federated Learning Google uses federated learning to train machine learning models on user devices without transferring raw data to central servers. This approach enhances privacy by keeping personal data on the device while still benefiting from large-scale data analysis.

Security Risks

Addressing Security Vulnerabilities and Potential for Misuse of AI: AI systems, like any other technology, are susceptible to security vulnerabilities. Ensuring the security of AI applications and preventing their misuse is essential to protect users and maintain trust in AI technologies.

- **Adversarial Attacks:**
  - **Definition:** Adversarial attacks involve manipulating input data to deceive AI models into making incorrect predictions. These attacks exploit the vulnerabilities in AI algorithms and can lead to serious consequences in applications such as autonomous driving, facial recognition, and cybersecurity.

- **Examples:** Adding subtle, imperceptible noise to images can cause a facial recognition system to misidentify individuals, or perturbing sensor data can mislead an autonomous vehicle.

Strategies to Mitigate Adversarial Attacks:

- **Robustness Testing:** Regularly testing AI models against adversarial examples to identify and fix vulnerabilities. This involves simulating various attack scenarios and strengthening the model's defenses.
- **Adversarial Training:** Incorporating adversarial examples into the training process to improve the model's resilience to attacks. By exposing the model to potential threats during training, it becomes better equipped to handle real-world adversarial inputs.

Example: Adversarial ML Threat Matrix Microsoft and MITRE have developed the Adversarial ML Threat Matrix, a framework for understanding and mitigating adversarial machine learning threats. It provides guidelines for assessing and defending against adversarial attacks in AI systems.

- **Misuse of AI:**
    - **Deepfakes:** AI-generated synthetic media, known as deepfakes, can create realistic but fake audio, video, and images. Deepfakes pose significant risks for misinformation, fraud, and privacy violations.
    - **Automated Cyberattacks:** AI can be used to automate and enhance cyberattacks, making them more sophisticated and difficult to detect. AI-powered malware and phishing attacks are examples of how AI can be misused for malicious purposes.

Strategies to Address AI Misuse:

- **Detection and Mitigation Tools:** Developing tools and algorithms to detect and mitigate the effects of deepfakes and other AI-generated synthetic media. These tools help identify fake content and prevent its spread.
- **Ethical Guidelines and Regulations:** Establishing ethical guidelines and regulatory frameworks to govern the use of AI technologies. These guidelines should address the potential misuse of AI and provide mechanisms for accountability and enforcement.

Example: Deepfake Detection Tools Researchers and companies are developing AI-based tools to detect deepfakes. These tools analyse visual and audio cues to identify inconsistencies and signs of manipulation, helping to verify the authenticity of media content.

Conclusion: Balancing AI innovation with privacy protection and addressing security vulnerabilities are critical to the responsible development and deployment of AI technologies. By implementing privacy-preserving techniques, enhancing the robustness of AI models, and establishing ethical guidelines and regulatory frameworks, we can mitigate the risks associated with AI while maximizing its benefits. Ensuring that AI systems are secure and trustworthy is essential for maintaining public confidence and fostering the continued advancement of AI.

This comprehensive exploration of privacy and security challenges in AI highlights the importance of protecting personal data, mitigating security vulnerabilities, and preventing the misuse of AI technologies. By examining specific examples and strategies, we can understand how to develop and deploy AI systems that are both innovative and ethically responsible.

2.6 Preparing for an AI-Driven Future

2.6.1 Education and Skills Development

The advent of artificial intelligence (AI) is reshaping numerous sectors, including education and the workforce. Ensuring that people are equipped with the necessary skills to thrive in an AI-driven economy is paramount. This section delves into the importance of AI literacy for all age groups and highlights programs and initiatives aimed at reskilling and upskilling the workforce.

AI Literacy

Importance of AI Education and Training for All Age Groups: AI literacy is crucial for enabling individuals to understand, interact with, and leverage AI technologies effectively. As AI becomes increasingly embedded in everyday life and various industries, a broad understanding of AI principles and applications is essential.

- **Early Education:**
    - **Introduction to AI Concepts:** Introducing AI concepts at an early age helps demystify the technology and encourages interest in STEM (Science, Technology, Engineering, and Mathematics) fields. Educational programs and curricula can integrate basic AI principles, such as pattern recognition, machine learning, and ethical considerations.
    - **Interactive Learning Tools:** AI-powered educational tools, such as coding games and virtual assistants, can make learning engaging and accessible. These tools can introduce children to programming and logical thinking in a fun and interactive manner.

Example: AI for Kids Organizations like AI4K12 provide resources and guidelines for integrating AI education into K-12 curricula. Their initiatives include lesson plans, activities, and tools designed to teach students about AI and its applications.

- **Higher Education:**

- **Specialized Courses and Degrees:** Universities and colleges are increasingly offering specialized courses and degree programs in AI and related fields. These programs cover topics such as machine learning, data science, robotics, and AI ethics.
- **Research Opportunities:** Higher education institutions provide opportunities for students to engage in AI research, contributing to advancements in the field and preparing them for careers in AI-driven industries.

Example: MIT's AI+Ethics Curriculum The Massachusetts Institute of Technology (MIT) offers a comprehensive AI+Ethics curriculum that combines technical AI training with courses on ethical and societal implications. This interdisciplinary approach ensures that students are well-rounded and prepared to address the complex challenges associated with AI.

- **Adult Learning and Lifelong Education:**
  - **Professional Development:** Continuous learning opportunities are essential for professionals to stay updated with the latest AI advancements. Workshops, online courses, and certifications in AI and data science can help professionals enhance their skills and remain competitive.
  - **Public Awareness Campaigns:** Initiatives aimed at raising public awareness about AI can help demystify the technology and promote informed decision-making. Public lectures, webinars, and media campaigns can play a significant role in educating the broader population about AI's benefits and risks.

Example: Coursera and edX AI Courses Platforms like Coursera and edX offer a wide range of AI courses from leading universities and institutions. These courses cater to different skill levels and cover various aspects of AI, from introductory overviews to advanced machine learning techniques.

Reskilling and Upskilling

Programs and Initiatives to Prepare the Workforce for an AI-Driven Economy: As AI technologies continue to evolve, there is a growing need for reskilling and upskilling programs to prepare the workforce for new roles and responsibilities. These initiatives aim to equip workers with the skills needed to thrive in an AI-driven economy.

- **Government and Public Sector Initiatives:**
  - **National AI Strategies:** Many countries are developing national AI strategies that include provisions for workforce development. These strategies often focus on funding education and training programs, fostering public-private partnerships, and promoting AI research and innovation.
  - **Public Training Programs:** Governments are launching training programs to help workers transition to new roles in AI-related fields. These programs often target sectors most affected by automation, such as manufacturing and customer service.

Example: Singapore's SkillsFuture Initiative Singapore's SkillsFuture initiative offers a range of training programs and subsidies to help individuals develop skills in emerging technologies, including AI. The initiative provides lifelong learning opportunities and encourages continuous skill development.

- **Corporate and Private Sector Initiatives:**
  - **In-House Training Programs:** Companies are investing in in-house training programs to reskill their employees for AI-related roles. These programs often include workshops, online courses, and mentorship opportunities.

- **Partnerships with Educational Institutions:** Collaborations between companies and educational institutions can create tailored training programs that address specific industry needs. These partnerships ensure that training is relevant and aligned with current market demands.

Example: Amazon's Upskilling 2025 Amazon's Upskilling 2025 program aims to provide training and education opportunities to 100,000 employees. The program focuses on areas such as data science, machine learning, and software engineering, preparing employees for higher-skilled roles within the company.

- **Non-Profit and Community-Based Initiatives:**
    - **Accessible Training Programs:** Non-profit organizations and community groups are developing training programs to ensure that underserved populations have access to AI education. These programs often focus on digital literacy, coding, and basic AI principles.
    - **Support Networks:** Building support networks and mentorship programs can help individuals navigate the challenges of reskilling and upskilling. These networks provide guidance, resources, and encouragement to learners at all stages of their careers.

Example: AI for Everyone by Coursera AI for Everyone, a course offered by Coursera and taught by Andrew Ng, aims to provide a broad understanding of AI concepts to non-technical audiences. The course is designed to be accessible to anyone, regardless of their background, and focuses on the impact of AI on society and businesses.

Conclusion: AI literacy and workforce development are crucial for preparing individuals and society for an AI-driven future. By investing in education and training programs across all age groups, we can ensure that people are equipped with the necessary skills to leverage AI technologies effectively. Reskilling and upskilling initiatives will help workers transition to new roles and contribute to a dynamic, AI-driven economy. Embracing lifelong learning and fostering collaboration between governments, educational institutions, and the private sector will be key to building a resilient and adaptable workforce.

This comprehensive exploration of education and skills development in the context of AI highlights the importance of AI literacy and the need for reskilling and upskilling programs. By examining specific examples and initiatives, we can understand how to prepare individuals and the workforce for the opportunities and challenges presented by AI.

2.6.2 Policy and Regulation

As artificial intelligence (AI) continues to evolve and integrate into various aspects of society, the development of effective policies and regulatory frameworks becomes crucial. These measures ensure that AI technologies are developed and deployed responsibly, ethically, and in a manner that benefits all stakeholders. This section explores the need for regulatory frameworks to govern AI and highlights the importance of global collaboration in addressing AI challenges and opportunities.

Regulatory Frameworks

Developing Policies and Regulations to Govern AI Development and Deployment: Regulatory frameworks are essential for ensuring that AI technologies are used responsibly and ethically. These frameworks address various aspects of AI development and deployment, including safety, fairness, transparency, accountability, and privacy.

- **Safety and Reliability:**
    - **Standards and Certification:** Establishing standards and certification processes for AI systems ensures that they meet specific safety and reliability criteria. These standards help prevent malfunctions and reduce the risk of harm to individuals and society.
    - **Risk Assessment:** Regulatory frameworks should include provisions for risk assessment and management. This involves identifying potential risks associated with AI systems and implementing measures to mitigate them.

Example: The European Commission's AI Act The European Commission has proposed the AI Act, which aims to create a comprehensive regulatory framework for AI in the European Union. The Act classifies AI systems based on their risk levels and sets out requirements for high-risk AI applications, including conformity assessments, transparency measures, and human oversight.

- **Fairness and Non-Discrimination:**
    - **Anti-Discrimination Policies:** Regulations should prohibit AI systems from perpetuating or exacerbating biases and discrimination. This involves ensuring that AI models are trained on diverse and representative datasets and implementing fairness checks throughout the development process.
    - **Equitable Access:** Policies should promote equitable access to AI technologies and their benefits. This includes addressing digital divides and ensuring that marginalized and underserved communities are not left behind.

Example: The Algorithmic Accountability Act In the United States, the Algorithmic Accountability Act requires companies to assess the impact of their automated decision-making systems, including AI, on fairness and discrimination. The Act mandates transparency and accountability for high-risk AI applications, aiming to prevent biased outcomes.

- **Transparency and Accountability:**
  - **Explainable AI:** Regulations should require AI systems to be transparent and explainable, ensuring that their decision-making processes can be understood by humans. This enhances accountability and builds trust in AI technologies.
  - **Accountability Mechanisms:** Establishing clear accountability mechanisms for AI developers and deployers is crucial. This includes defining responsibilities, providing avenues for redress, and implementing oversight and audit processes.

Example: The General Data Protection Regulation (GDPR) The GDPR includes provisions that impact AI, such as the right to explanation, which gives individuals the right to understand how automated decisions affecting them are made. The regulation emphasizes transparency, accountability, and data protection.

- **Privacy Protection:**
  - **Data Privacy Regulations:** Protecting individuals' privacy is a key concern in AI regulation. Policies should ensure that AI systems comply with data protection regulations, such as obtaining informed consent, minimizing data collection, and implementing robust data security measures.
  - **Privacy-Preserving Techniques:** Encouraging the use of privacy-preserving techniques, such as differential privacy and federated learning, can help balance the benefits of AI with the need to protect personal data.

Example: California Consumer Privacy Act (CCPA) The CCPA provides Californians with greater control over their personal data, including the right to know what data is collected and how it is used. The regulation impacts AI by ensuring that data-driven AI systems adhere to strict privacy standards.

Global Collaboration

Importance of International Cooperation in Addressing AI Challenges and Opportunities: AI's global impact necessitates international cooperation to address its challenges and harness its opportunities. Collaborative efforts can help create harmonized standards, share best practices, and ensure that AI benefits are distributed equitably across countries and regions.

- **Harmonized Standards and Guidelines:**
    - **International Standards:** Developing international standards for AI can help ensure consistency and interoperability across borders. Organizations such as the International Organization for Standardization (ISO) and the Institute of Electrical and Electronics Engineers (IEEE) play a crucial role in establishing these standards.
    - **Best Practices:** Sharing best practices and lessons learned from different countries and industries can enhance AI governance. International forums and conferences provide platforms for knowledge exchange and collaboration.

Example: The OECD AI Principles The Organisation for Economic Co-operation and Development (OECD) has developed AI principles that promote responsible stewardship of trustworthy AI. These principles provide a framework for governments to create policies that ensure AI systems are robust, safe, fair, and transparent.

- **Collaborative Research and Development:**
    - **Joint Research Initiatives:** International collaboration in AI research and development can accelerate innovation and address global challenges.

Joint initiatives can pool resources, expertise, and data to tackle complex problems.
- **Public-Private Partnerships:** Collaborations between governments, academia, and industry can drive advancements in AI while ensuring that ethical and societal considerations are addressed.

Example: The Partnership on AI The Partnership on AI is a global multi-stakeholder organization that brings together companies, non-profits, and academic institutions to collaborate on AI research and best practices. The partnership focuses on ensuring that AI is developed and used in ways that benefit society.

- **Addressing Global Challenges:**
    - **Sustainable Development Goals (SDGs):** AI has the potential to contribute to the United Nations' SDGs by addressing issues such as poverty, health, education, and climate change. International cooperation can align AI initiatives with these global goals.
    - **Ethical and Inclusive AI:** Ensuring that AI development is ethical and inclusive requires global collaboration. This includes addressing ethical dilemmas, mitigating biases, and ensuring that AI technologies are accessible to all.

Example: AI for Good Global Summit The AI for Good Global Summit, organized by the International Telecommunication Union (ITU) in partnership with other UN agencies, brings together AI experts and stakeholders to discuss how AI can be leveraged to achieve the SDGs. The summit promotes collaboration and the exchange of ideas to harness AI for societal good.

Conclusion: Effective policy and regulation are essential for governing AI development and deployment responsibly. By developing comprehensive regulatory frameworks and fostering global collaboration, we can address the ethical, societal, and technical challenges associated with AI. Ensuring that AI technologies are safe, fair, transparent, and accountable will help build public trust and maximize the benefits of AI for all. Collaborative efforts at the international level will be crucial in creating a cohesive and inclusive AI landscape that addresses global challenges and opportunities.

This comprehensive exploration of policy and regulation in the context of AI highlights the importance of developing robust regulatory frameworks and fostering international cooperation. By examining specific examples and initiatives, we can understand how to govern AI technologies responsibly and ensure that their benefits are shared globally.

2.6.3 AI for Social Good

Artificial intelligence (AI) holds immense potential for advancing social good by addressing some of the world's most pressing challenges. This section explores how AI can be leveraged in humanitarian efforts and sustainable development, illustrating its impact through case studies and practical examples.

AI in Humanitarian Efforts

Leveraging AI for Disaster Response: AI can significantly enhance disaster response efforts by providing timely and accurate information, optimizing resource allocation, and improving coordination among responders. Key applications include:

- **Predictive Analytics:** AI models can predict the occurrence and impact of natural disasters such as hurricanes, earthquakes, and floods. By analysing historical data and real-time information, these models help governments and organizations prepare and respond more effectively.

  - **Example: IBM's Watson for Disaster Management** IBM's Watson uses AI to predict natural disasters and assess their potential impacts. By analysing weather patterns, seismic activity, and other relevant data, Watson provides actionable insights that aid in disaster preparedness and response.

- **Damage Assessment:** AI-powered image recognition and satellite imagery analysis can quickly assess damage following a disaster. These tools help identify affected areas, prioritize response efforts, and allocate resources efficiently.

  - **Example: Google's Project Loon** Project Loon uses AI and machine learning to analyse satellite imagery and provide connectivity in disaster-stricken areas. By deploying high-altitude balloons, the project restores internet access, facilitating communication and coordination during disaster recovery.

- **Resource Allocation:** AI algorithms can optimize the distribution of aid and resources to disaster-affected areas. By analysing factors such as population density, infrastructure damage, and logistics, these algorithms ensure that aid reaches those who need it most.

- **Example: World Food Programme (WFP)** The WFP leverages AI to optimize food distribution in disaster zones. AI-driven models analyse data on food availability, transportation routes, and population needs to ensure efficient and effective delivery of aid.

Poverty Alleviation: AI can play a critical role in alleviating poverty by identifying vulnerable populations, optimizing social programs, and promoting economic development.

- **Targeted Interventions:** AI systems can analyse socioeconomic data to identify communities and individuals most in need of assistance. This enables governments and organizations to design and implement targeted poverty alleviation programs.

    - **Example: TATA Trusts' Data-Driven Initiatives** TATA Trusts in India uses AI to analyse demographic and economic data, identifying areas with high poverty rates. This information guides the design of targeted interventions, such as education and healthcare programs, to uplift marginalized communities.

- **Microfinance and Financial Inclusion:** AI-powered platforms can assess creditworthiness and provide microloans to underserved populations. By analysing non-traditional data sources, such as mobile phone usage and social media activity, these platforms enable access to financial services for those without traditional credit histories.

- **Example: Tala** Tala uses AI to offer microloans to individuals in developing countries. By analysing smartphone data, Tala assesses credit risk and provides instant loans, promoting financial inclusion and economic empowerment.

Healthcare Access: AI can improve healthcare access and outcomes by enhancing diagnostics, optimizing healthcare delivery, and supporting public health initiatives.

- **Telemedicine and Remote Diagnosis:** AI-powered telemedicine platforms enable remote diagnosis and consultation, making healthcare accessible to individuals in remote or underserved areas. These platforms use AI to analyse patient data and provide accurate medical advice.
  - **Example: Babylon Health** Babylon Health's AI-powered app offers remote consultations and diagnostics, providing medical advice based on patient symptoms and health records. The app improves access to healthcare for individuals in remote or underserved areas.
- **Disease Surveillance and Epidemic Response:** AI models can track the spread of diseases and predict outbreaks, supporting public health efforts to contain and manage epidemics. By analysing data from multiple sources, including social media and healthcare records, these models provide early warnings and actionable insights.
  - **Example: BlueDot** BlueDot uses AI to track and predict the spread of infectious diseases. By

analysing global data on air travel, climate, and disease outbreaks, BlueDot provides early warnings and insights that help public health authorities respond to epidemics.

Sustainable Development

Role of AI in Achieving Sustainable Development Goals (SDGs): AI has the potential to significantly contribute to the United Nations' Sustainable Development Goals (SDGs) by addressing challenges related to poverty, hunger, health, education, and environmental sustainability.

- **Zero Hunger (SDG 2):**
  - **Precision Agriculture:** AI-powered precision agriculture techniques optimize crop yields and reduce resource usage. By analysing data from sensors, satellites, and weather forecasts, AI systems provide farmers with actionable insights to improve productivity and sustainability.
    - **Example: IBM's Watson Decision Platform for Agriculture** IBM's Watson Decision Platform for Agriculture uses AI to analyse data on soil conditions, weather, and crop health. The platform provides farmers with recommendations to optimize planting, irrigation, and fertilization, enhancing food security and sustainability.

- **Good Health and Well-Being (SDG 3):**
  - **AI in Healthcare:** AI-driven tools improve healthcare delivery, enhance diagnostics, and

support public health initiatives. These tools help achieve better health outcomes and make healthcare more accessible and affordable.

- **Example: Google's AI for Breast Cancer Detection** Google's AI technology improves breast cancer detection by analysing mammograms with higher accuracy than human radiologists. This advancement enhances early detection and treatment, contributing to better health outcomes.

- **Quality Education (SDG 4):**
  - **Personalized Learning:** AI-powered educational platforms offer personalized learning experiences tailored to individual students' needs and abilities. These platforms improve educational outcomes and make quality education more accessible.
    - **Example: Duolingo** Duolingo uses AI to provide personalized language learning experiences. The platform adapts lessons based on the user's progress and learning style, making education more engaging and effective.

- **Climate Action (SDG 13):**
  - **Environmental Monitoring:** AI systems monitor environmental conditions, track deforestation, and predict the impacts of climate change. These

tools support conservation efforts and inform policies to mitigate environmental degradation.

- **Example: Microsoft's AI for Earth**
  Microsoft's AI for Earth initiative uses AI to address environmental challenges such as climate change, biodiversity conservation, and water management. The initiative supports projects that leverage AI to monitor and protect natural resources.

Conclusion: AI has the potential to drive significant social good by enhancing humanitarian efforts and supporting sustainable development. By leveraging AI technologies, we can address some of the world's most pressing challenges, from disaster response and poverty alleviation to healthcare access and environmental sustainability. Ensuring that AI is developed and deployed ethically and inclusively will be key to realizing its full potential for social good and achieving the Sustainable Development Goals.

This comprehensive exploration of AI for social good highlights its transformative potential in humanitarian efforts and sustainable development. By examining specific examples and initiatives, we can understand how AI can be harnessed to create a positive impact on society and contribute to a better, more sustainable future.

Chapter 3: Blockchain and Decentralization

3.1 Basics of Blockchain Technology

Blockchain technology, often described as the backbone of cryptocurrencies, is a revolutionary concept that has far-reaching implications beyond digital currencies. This section delves into the basics of blockchain technology, explaining its core concepts, historical background, workings, and different types of blockchains.

3.1.1 Introduction to Blockchain

Definition and Core Concepts: Blockchain is a decentralized, distributed ledger technology that records transactions across multiple computers in such a way that the registered transactions cannot be altered retroactively. This immutability ensures the integrity and transparency of the data. The core components of a blockchain include:

- **Blocks:** The fundamental units of a blockchain, each block contains a list of transactions. Blocks are linked together in a chronological sequence, forming a chain.
- **Chains:** A series of blocks linked together, where each block contains a reference (hash) to the previous block, creating a secure chain.
- **Nodes:** Independent computers that participate in the blockchain network by maintaining a copy of the entire blockchain. Nodes validate and relay transactions.

- **Distributed Ledger:** A database that is consensually shared and synchronized across multiple sites, institutions, or geographies. It allows transactions to have public "witnesses," thereby making a cyberattack more difficult.

Historical Background: The concept of blockchain technology was first introduced in 2008 by an anonymous person or group of people using the pseudonym Satoshi Nakamoto. In a whitepaper titled "Bitcoin: A Peer-to-Peer Electronic Cash System," Nakamoto outlined the principles of a decentralized digital currency. The first blockchain application, Bitcoin, was launched in 2009. Since then, blockchain technology has evolved and found applications in various sectors beyond cryptocurrencies, such as supply chain management, finance, healthcare, and more.

3.1.2 How Blockchain Works

Structure of a Block: A block in a blockchain is composed of several key components:

- **Block Header:** Contains metadata about the block, including the block's version, timestamp, a reference to the previous block's hash, the Merkle root (a hash of all transactions in the block), and a nonce (a number used once in cryptographic communication).
- **Timestamp:** Records the exact time when the block was created, ensuring the chronological order of the blockchain.
- **Transaction Data:** A list of transactions included in the block. Each transaction records the transfer of value or data between parties.

Hashing and Cryptography: Cryptographic hashing functions play a crucial role in securing the blockchain. A hash function takes an input (or 'message') and returns a fixed-size string of bytes. The output, called a hash value or digest, is unique to each input. Even a small change in the input drastically changes the output, which ensures data integrity.

- **SHA-256:** One of the most commonly used cryptographic hash functions in blockchain technology, particularly in Bitcoin. It generates a 256-bit hash value.
- **Cryptographic Security:** Hashing ensures that each block contains a unique identifier. The hash of each block's header includes the previous block's hash, creating a secure link between blocks.

Consensus Mechanisms: Consensus mechanisms are protocols used to achieve agreement on a single data value among distributed processes or systems. They ensure the reliability and security of the blockchain by validating transactions and blocks. Common consensus mechanisms include:

- **Proof of Work (PoW):** Used by Bitcoin and many other cryptocurrencies, PoW requires participants (miners) to solve complex mathematical puzzles to validate transactions and create new blocks. This process consumes significant computational power and energy.
- **Proof of Stake (PoS):** In PoS, validators are chosen based on the number of coins they hold and are willing to "stake" as collateral. This mechanism is considered more energy-efficient than PoW.
- **Delegated Proof of Stake (DPoS):** An evolution of PoS, DPoS allows stakeholders to vote for delegates

who will validate transactions and create blocks on their behalf. This approach aims to enhance scalability and efficiency.
- **Practical Byzantine Fault Tolerance (PBFT):** PBFT is designed to tolerate Byzantine faults, where nodes in the network may act maliciously or arbitrarily. It ensures consensus is reached even if some nodes are unreliable.

3.1.3 Types of Blockchains
Public vs. Private Blockchains:

- **Public Blockchains:** Also known as permissionless blockchains, these are open to anyone and operate in a fully decentralized manner. Anyone can participate as a node, validate transactions, and contribute to the network. Examples include Bitcoin and Ethereum.
    - **Advantages:** High transparency, security, and censorship resistance.
    - **Challenges:** Scalability issues and higher energy consumption due to consensus mechanisms like PoW.

- **Private Blockchains:** Also known as permissioned blockchains, these are restricted to a specific group of participants. Access is controlled, and only authorized entities can validate transactions and maintain the ledger. Examples include Hyperledger Fabric and R3 Corda.
    - **Advantages:** Improved scalability, faster transaction processing, and greater privacy.
    - **Challenges:** Reduced transparency and potential centralization of control.

Consortium Blockchains: Consortium or federated blockchains are a hybrid model that combines elements of both public and private blockchains. They are managed by a group of organizations rather than a single entity, offering a balance between decentralization and control.

- **Use Cases:** Consortium blockchains are ideal for industries where multiple organizations need to collaborate and share data securely, such as banking, supply chain management, and healthcare.
- **Example: Marco Polo Network** The Marco Polo Network is a consortium blockchain platform for trade finance. It allows banks and other financial institutions to collaborate securely, streamline processes, and reduce transaction costs.

Conclusion: Blockchain technology, with its decentralized and secure nature, offers transformative potential across various industries. Understanding the basics of blockchain, including its structure, cryptographic foundations, and different types of blockchains, is essential for grasping its implications and applications. As blockchain technology continues to evolve, it promises to play a pivotal role in reshaping the technological landscape and driving innovation in the digital economy.

This comprehensive exploration of blockchain technology provides a solid foundation for understanding its core concepts, historical background, and various implementations. By examining specific examples and mechanisms, readers can appreciate how blockchain is set to revolutionize industries and everyday life.

3.2 Cryptocurrencies and Digital Assets

Cryptocurrencies and digital assets have revolutionized the financial landscape, introducing new ways of conducting transactions, raising capital, and investing. This section explores the fundamentals of cryptocurrencies, how they work, and their economic implications through initial coin offerings and tokenomics.

3.2.1 Introduction to Cryptocurrencies

Bitcoin: Bitcoin, created by an anonymous individual or group known as Satoshi Nakamoto, was introduced in a 2008 whitepaper titled "Bitcoin: A Peer-to-Peer Electronic Cash System." Launched in 2009, Bitcoin was the first decentralized cryptocurrency, and it remains the most well-known and widely used digital currency.

- **Significance:**
    - **Decentralization:** Bitcoin operates without a central authority, relying on a peer-to-peer network to validate and record transactions.
    - **Digital Gold:** Often referred to as "digital gold," Bitcoin is seen as a store of value and a hedge against inflation due to its limited supply of 21 million coins.
    - **Innovation:** Bitcoin introduced the concept of blockchain technology, laying the foundation for subsequent cryptocurrencies and decentralized applications.

Altcoins: While Bitcoin paved the way, numerous other cryptocurrencies, known as altcoins, have emerged, each with unique features and use cases. Some of the significant altcoins include:

- **Ethereum (ETH):**

- **Smart Contracts:** Ethereum introduced smart contracts, self-executing contracts with the terms of the agreement directly written into code. This innovation enables decentralized applications (DApps) and decentralized finance (DeFi) platforms.
- **Ether:** The native cryptocurrency of the Ethereum network, Ether, is used to power transactions and computational services on the blockchain.

- **Ripple (XRP):**
  - **Cross-Border Payments:** Ripple focuses on enabling fast and cost-effective cross-border payments. Its consensus algorithm allows for rapid transaction settlement without the need for mining.
  - **Adoption:** Ripple's technology is used by numerous financial institutions to streamline international payments.

- **Litecoin (LTC):**
  - **Bitcoin Fork:** Created by Charlie Lee in 2011, Litecoin is a fork of Bitcoin with modifications such as a shorter block generation time and a different hashing algorithm (Scrypt).
  - **Faster Transactions:** Litecoin aims to provide faster transaction confirmation times and lower fees compared to Bitcoin.

- **Cardano (ADA):**
  - **Proof of Stake:** Cardano uses a unique proof-of-stake consensus mechanism called Ouroboros, which aims to be more energy-efficient than traditional proof-of-work systems.

- **Research-Driven:** Developed through peer-reviewed research, Cardano focuses on security, scalability, and interoperability.

- **Solana (SOL):**

    - **High Throughput:** Solana is known for its high throughput and low transaction fees, achieved through a combination of proof of stake and a novel consensus mechanism called Proof of History.
    - **DeFi and NFTs:** Solana has gained popularity for its DeFi applications and NFT marketplaces.

3.2.2 How Cryptocurrencies Work

Mining and Transactions: Cryptocurrency mining is the process by which new coins are created and transactions are validated on a blockchain network.

- **Proof of Work (PoW):**

    - **Mining Process:** In PoW systems like Bitcoin, miners compete to solve complex mathematical problems to validate transactions and add new blocks to the blockchain. The first miner to solve the problem is rewarded with newly created coins.
    - **Energy Consumption:** PoW requires significant computational power, leading to high energy consumption. This has raised concerns about the environmental impact of mining.

- **Proof of Stake (PoS):**

    - **Staking:** In PoS systems, validators are chosen based on the number of coins they hold and are willing to "stake" as collateral. Validators are selected to create new blocks and confirm transactions.

- **Energy Efficiency:** PoS is considered more energy-efficient than PoW as it does not rely on extensive computational power.

Wallets and Keys: Cryptocurrency wallets are digital tools that allow users to store, send, and receive cryptocurrencies. They use cryptographic keys to manage access and transactions.

- **Public and Private Keys:**
    - **Public Key:** The public key is a cryptographic code that allows users to receive cryptocurrency. It is similar to an account number and can be shared with others.
    - **Private Key:** The private key is a secure code that grants the owner access to their cryptocurrency. It must be kept confidential, as anyone with the private key can control the associated funds.

- **Types of Wallets:**
    - **Hot Wallets:** These are connected to the internet and offer convenience for frequent transactions. Examples include mobile wallets, web wallets, and desktop wallets.
    - **Cold Wallets:** These are offline wallets, providing enhanced security by keeping private keys offline. Examples include hardware wallets and paper wallets.

Example: Ledger Nano S The Ledger Nano S is a popular hardware wallet that provides secure storage for multiple cryptocurrencies. It keeps private keys offline and requires physical confirmation for transactions, enhancing security against online threats.

3.2.3 Initial Coin Offerings (ICOs) and Tokenomics

ICOs and STOs: Initial Coin Offerings (ICOs) and Security Token Offerings (STOs) are methods of fundraising for blockchain projects.

- **Initial Coin Offerings (ICOs):**
    - **Fundraising Mechanism:** ICOs allow companies to raise capital by issuing new cryptocurrencies or tokens to investors in exchange for established cryptocurrencies like Bitcoin or Ether.
    - **Utility Tokens:** Most tokens sold in ICOs are utility tokens, granting holders access to a product or service within the issuing company's ecosystem.

- **Security Token Offerings (STOs):**
    - **Regulated Fundraising:** STOs involve the issuance of security tokens that represent ownership in an asset, such as shares in a company or real estate. STOs are subject to regulatory oversight and must comply with securities laws.
    - **Investor Protection:** STOs aim to provide greater investor protection compared to ICOs, as they are backed by tangible assets and regulated by financial authorities.

Example: Ethereum's ICO Ethereum's ICO in 2014 raised over $18 million, making it one of the most successful ICOs in history. The funds were used to develop the Ethereum platform, which has since become a cornerstone of the blockchain ecosystem.

Token Economics: Tokenomics refers to the study of the economics of cryptocurrencies and tokens within a blockchain ecosystem. It includes the design, issuance, and management of tokens to create sustainable and efficient blockchain economies.

- **Utility Tokens:** Utility tokens provide access to a product or service within a blockchain ecosystem. They incentivize user participation and drive network effects.
- **Security Tokens:** Security tokens represent ownership in an asset or company. They offer potential returns through dividends or asset appreciation.
- **Governance Tokens:** These tokens grant holders voting rights on protocol changes and project governance decisions. They are used in decentralized autonomous organizations (DAOs) to facilitate community-driven decision-making.

Example: MakerDAO's MKR Token MakerDAO uses the MKR token for governance and stability within its decentralized finance (DeFi) platform. MKR holders vote on proposals related to the platform's development and risk management, playing a crucial role in its operation.

Conclusion: Cryptocurrencies and digital assets have introduced a new paradigm in finance, enabling decentralized transactions, innovative fundraising mechanisms, and the creation of digital economies. Understanding the fundamentals of cryptocurrencies, how they work, and their economic implications through ICOs and tokenomics is essential for navigating this rapidly evolving landscape. As these technologies continue to mature, they will likely play an increasingly significant role in reshaping industries and everyday life.

This comprehensive exploration of cryptocurrencies and digital assets provides a solid foundation for understanding their core concepts, workings, and economic implications. By examining specific examples and mechanisms, readers can appreciate how these technologies are revolutionizing the financial landscape and beyond.

3.3 Blockchain in Finance

Blockchain technology is revolutionizing the finance sector by introducing new paradigms for conducting financial transactions and managing assets. This section explores the impact of blockchain on finance, focusing on decentralized finance (DeFi), its disruption of traditional banking, and the regulatory landscape.

3.3.1 Decentralized Finance (DeFi)

Overview of DeFi: Decentralized Finance (DeFi) refers to a financial ecosystem built on blockchain technology that operates without traditional intermediaries like banks and financial institutions. DeFi leverages smart contracts on blockchain networks, primarily Ethereum, to offer a range of financial services in a decentralized manner. Key components of DeFi include:

- **Decentralized Exchanges (DEXs):**
    - **Functionality:** DEXs facilitate the direct exchange of cryptocurrencies between users without the need for an intermediary. They operate through smart contracts that automatically match buy and sell orders.
    - **Advantages:** DEXs offer increased privacy, security, and control over assets, as users retain custody of their funds. Examples include Uniswap, SushiSwap, and Balancer.

- **Lending Platforms:**
    - **Functionality:** DeFi lending platforms enable users to lend and borrow cryptocurrencies in a trustless environment. Lenders earn interest on their assets, while borrowers provide collateral to secure loans.
    - **Advantages:** These platforms offer competitive interest rates, global accessibility, and instant loan approvals. Examples include Aave, Compound, and MakerDAO.

- **Stablecoins:**
    - **Definition:** Stablecoins are cryptocurrencies pegged to stable assets like fiat currencies or commodities. They provide stability in value, making them suitable for everyday transactions and as a store of value.
    - **Types:** There are fiat-collateralized stablecoins (e.g., USDC, USDT), crypto-collateralized stablecoins (e.g., DAI), and algorithmic stablecoins (e.g., TerraUSD).

Smart Contracts: Smart contracts are self-executing contracts with the terms of the agreement directly written into code. They automate and enforce the execution of financial transactions on the blockchain, ensuring that transactions are transparent, tamper-proof, and efficient.

- **How They Work:** Smart contracts run on blockchain networks and automatically execute actions when predefined conditions are met. For example, a smart contract for a DeFi loan might release collateral to the lender if the borrower fails to repay the loan on time.
- **Role in DeFi:** Smart contracts are the backbone of DeFi applications, enabling functionalities such as trading on DEXs, automated lending and borrowing, yield farming, and staking.

Example: MakerDAO and DAI Stablecoin MakerDAO is a DeFi platform that uses smart contracts to create and manage the DAI stablecoin. Users can lock collateral (such as ETH) in smart contracts to generate DAI, which is pegged to the US dollar. The system is governed by MKR token holders who vote on key decisions.

3.3.2 Impact on Traditional Banking
Banking Disruption: Blockchain technology is disrupting traditional banking and financial services by offering more efficient,

secure, and transparent alternatives to conventional financial systems.

- **Disintermediation:** Blockchain removes the need for intermediaries like banks, reducing transaction costs and increasing the speed of financial services.
- **Transparency:** Transactions on the blockchain are transparent and immutable, enhancing trust and reducing the potential for fraud.
- **Financial Inclusion:** Blockchain and DeFi platforms provide financial services to the unbanked and underbanked populations, offering access to loans, savings, and investments without the need for a traditional bank account.

Example: Decentralized Lending on Aave Aave is a DeFi lending platform that allows users to lend and borrow a wide range of cryptocurrencies without intermediaries. Lenders earn interest on their deposits, while borrowers provide collateral to secure loans, all facilitated by smart contracts.

Cross-Border Payments: Blockchain technology significantly improves cross-border payments and remittances by reducing costs, increasing speed, and enhancing transparency.

- **Reduced Costs:** Traditional cross-border payments involve multiple intermediaries and high fees. Blockchain enables direct peer-to-peer transactions, reducing fees associated with intermediaries.
- **Speed:** Blockchain transactions are processed in minutes, compared to several days for traditional cross-border transfers.
- **Transparency and Security:** Each transaction is recorded on the blockchain, providing a transparent and immutable record. This reduces the risk of fraud and enhances security.

Example: Ripple (XRP) Ripple is a blockchain-based payment protocol designed for fast and cost-effective cross-border transactions. Ripple's technology is used by financial institutions to streamline international payments, offering near-instant settlement times and lower transaction costs.

3.3.3 Regulatory Landscape

Regulatory Challenges: The rapid growth of blockchain and cryptocurrencies has posed significant regulatory challenges. Governments and regulatory bodies worldwide are grappling with how to oversee these technologies effectively while fostering innovation.

- **Jurisdictional Differences:** Regulatory approaches vary widely across different countries, creating challenges for global blockchain projects. Some countries have embraced blockchain technology, while others have imposed strict regulations or outright bans.
- **Consumer Protection:** Regulators aim to protect consumers from fraud, scams, and market manipulation in the largely unregulated cryptocurrency space.
- **Financial Stability:** There are concerns about the impact of cryptocurrencies on financial stability, particularly regarding their potential use for money laundering and other illicit activities.

Example: Regulatory Approaches in the US and EU In the United States, regulatory agencies like the SEC, CFTC, and FinCEN oversee different aspects of the cryptocurrency market. The European Union is developing the Markets in Crypto-Assets (MiCA) regulation to provide a comprehensive framework for crypto-assets and their service providers.

Compliance and Legal Issues: Compliance with regulatory requirements such as Know Your Customer (KYC) and Anti-Money Laundering (AML) regulations is crucial for blockchain and cryptocurrency projects to operate legally and gain legitimacy.

- **Know Your Customer (KYC):** KYC regulations require financial institutions to verify the identity of their customers to prevent fraud and money laundering. DeFi platforms and

exchanges are increasingly implementing KYC procedures to comply with regulations.
- **Anti-Money Laundering (AML):** AML regulations aim to prevent the use of financial systems for money laundering and terrorist financing. Blockchain analytics tools help trace the flow of funds and identify suspicious activities.

Example: Chainalysis Chainalysis is a blockchain analytics company that provides compliance and investigation solutions to help businesses and governments detect and prevent cryptocurrency-related crime. Their tools are used by exchanges, financial institutions, and law enforcement agencies.

Conclusion: Blockchain technology is transforming the financial sector by introducing decentralized finance, disrupting traditional banking, and enhancing cross-border payments. However, the growth of blockchain and cryptocurrencies also brings regulatory challenges that must be addressed to ensure consumer protection, financial stability, and compliance with legal standards. As regulatory frameworks evolve, they will play a crucial role in shaping the future of blockchain in finance and ensuring its responsible development and adoption.

This comprehensive exploration of blockchain in finance highlights its transformative impact on decentralized finance, traditional banking, and the regulatory landscape. By examining specific examples and mechanisms, readers can appreciate how blockchain technology is reshaping the financial sector and the challenges that must be addressed to realize its full potential.

3.4 Blockchain in Supply Chain Management

Blockchain technology is transforming supply chain management by enhancing transparency, efficiency, and trust.

This section explores how blockchain can improve supply chain transparency and traceability, streamline processes to reduce costs, and examines real-world case studies of successful implementations.

3.4.1 Supply Chain Transparency

Tracking and Traceability: Blockchain technology offers unprecedented transparency and traceability in supply chains by providing a decentralized and immutable ledger for recording transactions. This capability is particularly valuable in industries where the provenance and safety of products are critical.

- **Enhanced Visibility:** Each participant in the supply chain can access a single, shared version of the truth. This real-time visibility into every transaction and movement of goods ensures that all stakeholders are informed and accountable.
- **End-to-End Traceability:** Blockchain enables end-to-end traceability from raw materials to finished products. This comprehensive tracking helps identify the origin of components, monitor their journey, and ensure compliance with regulatory requirements.

Example: IBM Food Trust IBM Food Trust uses blockchain to improve food safety and traceability. By tracking the journey of food products from farm to table, the platform helps identify contamination sources quickly, reducing the risk of widespread foodborne illnesses and enhancing consumer confidence.

Provenance and Authenticity: Blockchain ensures the authenticity and provenance of products by providing tamper-proof records that verify each step in the supply chain.

- **Counterfeit Prevention:** Blockchain's immutable ledger makes it nearly impossible to alter records, helping to prevent counterfeit goods from entering the supply chain. This is crucial in industries like pharmaceuticals, luxury goods, and electronics.
- **Verification of Authenticity:** Consumers and businesses can verify the authenticity of products by scanning QR codes or using blockchain-based verification tools. This transparency builds trust and supports ethical sourcing practices.

Example: Everledger Everledger uses blockchain to track the provenance of diamonds, providing a digital certificate of authenticity that includes information on the diamond's origin, characteristics, and ownership history. This helps combat fraud and ensures ethical sourcing.

3.4.2 Efficiency and Cost Reduction

Streamlining Processes: Blockchain streamlines supply chain processes by reducing the need for intermediaries, automating transactions, and improving data accuracy.

- **Reduced Intermediaries:** By enabling direct peer-to-peer transactions, blockchain reduces reliance on intermediaries such as brokers and agents. This not only lowers costs but also speeds up transactions and reduces the potential for errors.
- **Improved Data Accuracy:** Blockchain's decentralized nature ensures that all participants have access to the same data, reducing discrepancies and the need for reconciliations. This improves data accuracy and consistency across the supply chain.

Example: TradeLens TradeLens, a blockchain platform developed by Maersk and IBM, digitizes the global supply chain by connecting all stakeholders on a single platform. This reduces paperwork, speeds up customs processing, and enhances transparency, leading to significant cost savings.

Smart Contracts in Supply Chains: Smart contracts automate and enforce agreements in supply chain transactions, ensuring that terms are met without the need for manual intervention.

- **Automation of Transactions:** Smart contracts execute automatically when predefined conditions are met. For example, payment can be released to a supplier once goods are delivered and verified, without the need for human intervention.
- **Enforcement of Agreements:** Smart contracts ensure that all parties adhere to the terms of the agreement. This reduces disputes and the need for legal proceedings, saving time and costs.

Example: SkuChain SkuChain uses blockchain and smart contracts to create a collaborative commerce platform for supply chains. Smart contracts automate transactions and enforce agreements, enhancing efficiency and reducing the risk of disputes.

3.4.3 Case Studies

Real-World Examples: Several companies have successfully implemented blockchain in their supply chains, demonstrating its potential to enhance transparency, efficiency, and trust.

- **Walmart:** Walmart has partnered with IBM to use blockchain for tracking leafy greens. By recording every step of the supply chain on a blockchain, Walmart can trace the origin of products in seconds, improving food safety and reducing the time needed to respond to contamination issues.

- **IBM Food Trust:** IBM Food Trust is a blockchain-based platform that brings together participants across the food supply chain, including farmers, processors, distributors, and retailers. The platform enhances traceability, reduces waste, and ensures food safety by providing a transparent and tamper-proof record of each transaction.

- **Maersk's TradeLens:** TradeLens is a blockchain-based shipping platform developed by Maersk and IBM. It connects all parties in the global supply chain, including shippers, ports, customs authorities, and logistics providers. TradeLens improves transparency, reduces paperwork, and speeds up customs processing, leading to significant cost savings and operational efficiencies.

Example: Provenance Provenance is a blockchain-based platform that tracks the journey of products from source to consumer. It provides transparency and proof of authenticity, helping brands build trust with consumers by ensuring ethical and sustainable sourcing practices.

Conclusion: Blockchain technology is revolutionizing supply chain management by providing enhanced transparency, traceability, and efficiency. By reducing intermediaries, automating transactions, and ensuring data accuracy, blockchain helps streamline supply chain processes and reduce costs. Real-world case studies demonstrate the significant benefits of blockchain implementation in supply chains, highlighting its potential to transform industries and enhance consumer trust. As blockchain technology continues to evolve, it will play an increasingly important role in creating more efficient, secure, and transparent supply chains.

This comprehensive exploration of blockchain in supply chain management highlights its transformative impact on transparency, efficiency, and trust. By examining specific examples and mechanisms, readers can appreciate how blockchain technology is reshaping supply chains and the benefits it brings to businesses and consumers alike.

3.5 Blockchain and Data Security

Blockchain technology offers significant advancements in data security, providing immutable records, decentralized storage, and robust privacy measures. This section explores how blockchain enhances data security, addresses privacy concerns, and contributes to cybersecurity solutions.

3.5.1 Enhanced Security

Immutable Records: One of the foundational attributes of blockchain technology is its immutability. Once data is recorded on the blockchain, it cannot be altered or deleted. This characteristic is crucial for enhancing data security.

- **Tamper-Proof Ledger:** The immutability of blockchain ensures that all transactions are permanently recorded and visible to all network participants. This transparency and permanence make it extremely difficult for malicious actors to alter historical data.

- **Trust and Integrity:** Immutable records foster trust among users, as they can be confident that the data has not been tampered with. This is particularly important in sectors like finance, healthcare, and supply chain management, where data integrity is critical.

Example: Financial Audits Blockchain's immutability is beneficial for financial audits, providing a transparent and verifiable trail of all transactions. Auditors can access an unalterable record, reducing the risk of fraud and improving the accuracy of financial statements.

Decentralized Storage: Decentralized storage solutions leverage blockchain technology to enhance data security and redundancy, mitigating the risks associated with centralized data storage.

- **Data Redundancy:** Decentralized storage distributes data across multiple nodes in the network. This redundancy ensures that data remains accessible even if some nodes fail or are compromised.
- **Reduced Single Point of Failure:** Traditional centralized storage systems are vulnerable to attacks and failures. Decentralized storage eliminates single points of failure, enhancing overall data security and resilience.

Example: Filecoin Filecoin is a decentralized storage network that incentivizes users to share their unused storage space. By distributing data across a decentralized network, Filecoin enhances data security and availability, making it less susceptible to attacks and failures.

3.5.2 Privacy Concerns
Anonymity vs. Privacy: Blockchain technology offers varying degrees of anonymity and privacy, and finding the right balance is crucial for different use cases.

- **Anonymity:** Public blockchains like Bitcoin provide pseudonymity, where users are identified by alphanumeric

addresses rather than personal information. While this offers a degree of anonymity, transactions are publicly visible and can potentially be traced.
- **Privacy:** Achieving privacy involves ensuring that transaction details and user identities are protected from public view. Privacy-enhancing technologies like zero-knowledge proofs (ZKPs) are essential for maintaining confidentiality while enabling verification.

Example: Monero Monero is a cryptocurrency that emphasizes privacy and anonymity. It uses advanced cryptographic techniques to obfuscate transaction details, making it difficult to trace transactions and link them to specific users.

Zero-Knowledge Proofs: Zero-knowledge proofs (ZKPs) are cryptographic methods that enable one party to prove to another that a statement is true without revealing any information beyond the validity of the statement itself.

- **Enhanced Privacy:** ZKPs allow users to verify transactions without exposing transaction details. This enhances privacy while maintaining the integrity and security of the blockchain.
- **Applications:** ZKPs are used in various blockchain applications, including confidential transactions, identity verification, and secure voting systems.

Example: zk-SNARKs zk-SNARKs (Zero-Knowledge Succinct Non-Interactive Arguments of Knowledge) are a type of ZKP used by cryptocurrencies like Zcash to enable private transactions. They allow users to prove the validity of a transaction without revealing the transaction amount or participants.

3.5.3 Blockchain in Cybersecurity
Security Protocols: Blockchain technology can enhance security protocols, providing robust defense mechanisms against cyber attacks.

- **Distributed Consensus:** Blockchain's consensus mechanisms (e.g., proof of work, proof of stake) ensure that the majority of network participants agree on the validity of transactions. This decentralized approach makes it challenging for attackers to compromise the network.
- **Tamper-Resistance:** The cryptographic nature of blockchain transactions makes it difficult for unauthorized parties to alter or delete data, protecting against tampering and fraud.

Example: Guardtime Guardtime uses blockchain technology to secure data integrity for various industries, including defense, healthcare, and finance. Their Keyless Signature Infrastructure (KSI) provides tamper-evident and cryptographically verifiable signatures for digital assets, enhancing cybersecurity.

Identity Management: Blockchain technology offers innovative solutions for secure identity management and authentication systems, addressing issues like identity theft and data breaches.

- **Decentralized Identity (DID):** Decentralized identity solutions use blockchain to give individuals control over their digital identities. Users can manage and share their identity attributes securely without relying on centralized authorities.
- **Secure Authentication:** Blockchain-based authentication systems use cryptographic keys for secure access control. This reduces the risk of password breaches and enhances the security of digital interactions.

Example: Sovrin Sovrin is a decentralized identity network that uses blockchain technology to create self-sovereign identities. Users can manage their identity data and share verified credentials with third parties securely and privately.

Conclusion: Blockchain technology offers robust solutions for enhancing data security, addressing privacy concerns, and improving cybersecurity. By leveraging immutable records, decentralized storage, zero-knowledge proofs, and secure identity management, blockchain provides a secure and transparent framework for managing data and transactions. As these technologies continue to evolve, they will play a critical role in protecting digital assets and ensuring the integrity of information in an increasingly connected world.

This comprehensive exploration of blockchain and data security highlights its transformative potential in enhancing security, addressing privacy concerns, and contributing to cybersecurity. By examining specific examples and mechanisms, readers can appreciate how blockchain technology is reshaping the landscape of data security and privacy.

3.6 Challenges and Future Directions

As blockchain technology continues to mature, it faces several challenges that need to be addressed to realize its full potential. This section explores the key challenges related to scalability, energy consumption, and interoperability, and discusses future trends and directions for blockchain technology.

3.6.1 Scalability Issues

Current Limitations: Scalability is one of the most significant challenges facing current blockchain networks. As the number of users and transactions grows, the network's ability to process transactions quickly and efficiently is strained.

- **Throughput:** Blockchain networks like Bitcoin and Ethereum have limited throughput, processing only a handful of transactions per second compared to traditional payment systems like Visa, which can handle thousands.
- **Latency:** High transaction volumes can lead to network congestion and increased latency, causing delays in transaction confirmation times.

- **Resource Requirements:** As blockchain networks scale, the computational and storage requirements for nodes increase, potentially leading to centralization as fewer participants can afford to maintain full nodes.

Solutions and Innovations: Several solutions and innovations are being developed to address scalability issues and improve blockchain performance.

- **Sharding:** Sharding involves partitioning the blockchain network into smaller, more manageable segments called shards. Each shard processes a subset of transactions, allowing the network to handle more transactions in parallel.
    - **Example: Ethereum 2.0:** Ethereum 2.0, also known as Serenity, plans to implement sharding to increase its transaction throughput and reduce latency.

- **Sidechains:** Sidechains are separate blockchains that run parallel to the main chain (parent blockchain) and are connected to it via a two-way peg. Sidechains can process transactions independently, reducing the load on the main chain.
    - **Example: Liquid Network:** Liquid Network is a sidechain of Bitcoin designed to facilitate faster and more efficient transactions for exchanges and traders.

- **Layer 2 Solutions:** Layer 2 solutions build on top of the main blockchain to offload some transaction processing, improving scalability and reducing congestion.
    - **Lightning Network:** The Lightning Network is a Layer 2 solution for Bitcoin that enables fast and low-cost micropayments by creating off-chain payment channels.
    - **Plasma:** Plasma is a Layer 2 scaling solution for Ethereum that involves creating child chains that can

process transactions independently of the main Ethereum chain.

3.6.2 Energy Consumption

Environmental Impact: Energy consumption is a critical concern for blockchain networks that use energy-intensive consensus mechanisms like Proof of Work (PoW).

- **Proof of Work:** PoW requires miners to solve complex mathematical puzzles to validate transactions and secure the network. This process consumes significant amounts of electricity, contributing to environmental degradation.
- **Carbon Footprint:** The energy consumption associated with PoW mining contributes to a large carbon footprint, raising concerns about the sustainability of blockchain technology.

Sustainable Alternatives: To mitigate the environmental impact, researchers and developers are exploring more sustainable consensus mechanisms and energy-efficient solutions.

- **Proof of Stake (PoS):** PoS reduces energy consumption by selecting validators based on the number of tokens they hold and are willing to "stake" as collateral. This approach requires significantly less computational power compared to PoW.
    - **Example: Ethereum 2.0:** Ethereum 2.0 is transitioning from PoW to PoS to improve energy efficiency and scalability.

- **Other Energy-Efficient Algorithms:** Alternative consensus algorithms such as Delegated Proof of Stake (DPoS), Practical Byzantine Fault Tolerance (PBFT), and Proof of Authority (PoA) offer more energy-efficient solutions.
    - **Example: EOS:** EOS uses DPoS, where token holders vote for a limited number of delegates who validate transactions, reducing energy consumption compared to PoW.

3.6.3 Interoperability

Cross-Chain Communication: Achieving interoperability between different blockchain networks is crucial for the seamless exchange of information and assets.

- **Challenges:** Different blockchain networks often operate in isolation, with varying protocols, consensus mechanisms, and data structures, making cross-chain communication challenging.
- **Solutions:** Solutions for cross-chain interoperability include atomic swaps, blockchain bridges, and interoperable protocols.
    - **Atomic Swaps:** Atomic swaps enable direct peer-to-peer exchanges of cryptocurrencies between different blockchains without the need for intermediaries.
    - **Blockchain Bridges:** Blockchain bridges connect separate blockchain networks, allowing assets and information to move between them.
    - **Interoperable Protocols:** Protocols like Polkadot and Cosmos facilitate interoperability by providing a framework for connecting multiple blockchains.

Example: Polkadot Polkadot is an interoperable blockchain platform that connects multiple blockchains (parachains) to enable cross-chain communication and asset transfers. Polkadot's relay chain ensures security and consensus across all connected parachains.

Standardization Efforts: Standardization efforts are essential for achieving interoperability and ensuring that different blockchain networks can work together seamlessly.

- **Enterprise Ethereum Alliance (EEA):** The EEA is a consortium of organizations working to develop standards and specifications for enterprise use of Ethereum. Their

efforts promote interoperability and compatibility between Ethereum-based solutions.
- **Hyperledger Project:** Hyperledger is an open-source collaborative effort hosted by the Linux Foundation, focusing on developing cross-industry blockchain technologies. Hyperledger projects like Fabric and Sawtooth aim to provide interoperable blockchain frameworks for various use cases.

3.6.4 Future Trends

Evolving Use Cases: Blockchain technology is continually evolving, with emerging use cases extending beyond finance and supply chain management to other sectors.

- **Healthcare:** Blockchain can improve healthcare by providing secure and interoperable health records, enhancing data privacy, and enabling transparent clinical trials.
  - **Example: Medicalchain:** Medicalchain uses blockchain to create secure, patient-centric health records that can be shared with healthcare providers while maintaining patient privacy.

- **Voting Systems:** Blockchain-based voting systems offer secure, transparent, and tamper-proof election processes, reducing the risk of fraud and increasing voter trust.
  - **Example: Voatz:** Voatz is a mobile voting platform that uses blockchain to ensure secure and transparent voting, particularly for overseas and military voters.

- **Digital Identity:** Blockchain can provide secure and self-sovereign digital identity solutions, giving individuals control over their personal data and reducing identity theft.
  - **Example: Sovrin:** Sovrin is a decentralized identity network that uses blockchain to create self-sovereign identities, allowing individuals to manage and share their identity information securely.

Integration with Other Technologies: The integration of blockchain with other cutting-edge technologies like artificial intelligence (AI), the Internet of Things (IoT), and quantum computing holds significant potential for creating innovative solutions.

- **AI and Blockchain:** Combining AI with blockchain can enhance data security, transparency, and trust in AI models. Blockchain can provide immutable audit trails for AI decisions, improving accountability.
    - **Example: SingularityNET:** SingularityNET is a decentralized AI network that uses blockchain to create a marketplace for AI services, enabling secure and transparent collaboration between AI developers and users.

- **IoT and Blockchain:** Integrating blockchain with IoT can enhance the security and interoperability of IoT devices, ensuring secure data exchange and reducing the risk of cyber attacks.
    - **Example: IOTA:** IOTA is a blockchain-based platform designed for IoT applications, providing secure data transfer and microtransactions between IoT devices.

- **Quantum Computing and Blockchain:** Quantum computing has the potential to break current cryptographic algorithms used in blockchain. Research is ongoing to develop quantum-resistant cryptographic techniques to secure blockchain networks against future quantum threats.
    - **Example: Quantum-Resistant Ledger (QRL):** QRL is a blockchain platform that uses quantum-resistant cryptographic algorithms to ensure security against potential quantum computing attacks.

Conclusion: While blockchain technology faces significant challenges related to scalability, energy consumption, and interoperability, ongoing innovations and research are paving the way for future advancements. Emerging use cases and the integration of blockchain with AI, IoT, and quantum computing hold promise for creating innovative and transformative solutions. As blockchain continues to evolve, it will play a crucial role in reshaping industries and enhancing everyday life in the coming years.

This comprehensive exploration of challenges and future directions in blockchain technology highlights the ongoing efforts to address scalability, energy consumption, and interoperability issues. By examining future trends and potential integrations with other technologies, readers can appreciate the transformative potential of blockchain in various sectors and its impact on the future of technology and society.

•

Chapter 4: Convergence of Technologies

4. *4.1 How Quantum Computing Will Enhance AI Capabilities*

Quantum computing represents a paradigm shift in computational power, promising to solve problems that are currently intractable for classical computers. When combined with artificial intelligence (AI), quantum computing has the potential to revolutionize the field, enhancing capabilities in unprecedented ways. This section explores how quantum computing will enhance AI, focusing on speed, efficiency, complex problem solving, advancements in neural networks, and real-world applications.

4.1.1 Introduction to Quantum-Enhanced AI

***Synergy Between Quantum* Computing and AI:** Quantum computing and AI are two of the most transformative technologies of our time. When integrated, they offer a synergistic relationship where quantum computing can significantly enhance AI by solving complex problems more efficiently and effectively.

- **Revolutionizing AI:** Quantum computing's ability to process vast amounts of data at unparalleled speeds and perform complex calculations that are infeasible for classical computers opens new avenues for AI development. This includes improving machine learning algorithms, optimizing neural networks, and enhancing data analysis capabilities.
- **Breaking Barriers:** Quantum-enhanced AI can break through current computational limitations, enabling more accurate predictions, better decision-making processes, and faster insights in various domains.

Quantum Machine Learning (QML): Quantum machine learning (QML) is an emerging field that combines quantum computing with machine learning techniques. QML leverages quantum algorithms to improve the efficiency and performance of machine learning models.

- **Potential Benefits:** QML can accelerate the training of machine learning models, enhance pattern recognition, and enable the processing of complex datasets that classical algorithms struggle with. This leads to more robust and accurate AI systems.

4.1.2 Speed and Efficiency

Quantum Algorithms for AI: Quantum algorithms have the potential to dramatically accelerate AI tasks, offering significant speed-ups compared to classical algorithms.

- **Quantum Annealing:** Quantum annealing is used for solving optimization problems by finding the global minimum of a function. It can be applied to machine learning tasks such as feature selection and hyperparameter tuning.

    - **Example:** D-Wave's quantum annealers have been used to optimize traffic flow, improving efficiency and reducing congestion in urban areas.

- **Quantum Support Vector Machines (QSVMs):** QSVMs are quantum versions of classical support vector machines, used for classification and regression tasks. They can process data much faster and more accurately than classical SVMs.

    - **Example:** QSVMs can be used in image recognition tasks, enhancing the accuracy and speed of detecting objects in large datasets.

Parallel Processing: Quantum computers can perform parallel processing on a massive scale, vastly increasing the speed of machine learning model training.

- **Superposition and Entanglement:** Quantum bits (qubits) leverage the principles of superposition and entanglement, allowing them to represent and process multiple states simultaneously. This enables quantum computers to evaluate numerous possibilities in parallel, significantly speeding up computations.
- **Accelerated Training:** The parallel processing capability of quantum computers can reduce the time required to train machine learning models, enabling faster deployment and iteration of AI systems.

4.1.3 Complex Problem Solving

Optimization Problems: Quantum computing excels at solving complex optimization problems, which are often computationally intensive and challenging for classical computers.

- **Combinatorial Optimization:** Problems such as the traveling salesman problem, portfolio optimization, and scheduling are examples where quantum computing can provide optimal or near-optimal solutions more efficiently.
  - **Example:** Quantum computing can optimize supply chain logistics, reducing costs and improving delivery times by finding the most efficient routes and schedules.

Large-Scale Data Analysis: Quantum computing's ability to process and analyse massive datasets can uncover patterns and insights that are beyond the reach of classical methods.

- **Pattern Recognition:** Quantum algorithms can enhance pattern recognition in large datasets, improving the accuracy of AI models in tasks such as anomaly detection, predictive analytics, and natural language processing.
  - **Example:** Quantum-enhanced AI can be used in cybersecurity to detect and mitigate threats by analysing vast amounts of network traffic data in real-time.

4.1.4 Advancements in Neural Networks

Quantum Neural Networks (QNNs): Quantum neural networks (QNNs) are a novel approach that leverages quantum computing to enhance the capabilities of traditional neural networks.

- **Outperforming Classical Networks:** QNNs have the potential to outperform classical neural networks in terms of speed and accuracy, particularly in tasks involving high-dimensional data and complex pattern recognition.

- **Example:** QNNs can be used in image and speech recognition tasks, providing faster and more accurate results compared to classical neural networks.

Hybrid Classical-Quantum Models: The integration of classical and quantum computing can further enhance the performance of AI models, combining the strengths of both paradigms.

- **Best of Both Worlds:** Hybrid models use classical computers for tasks they excel at, such as data preprocessing and simple computations, while leveraging quantum computers for complex, computationally intensive tasks.
  - **Example:** In drug discovery, classical computers can handle initial data analysis and hypothesis generation, while quantum computers can simulate molecular interactions and optimize drug candidates.

4.1.5 Real-World Applications

Healthcare and Drug Discovery: Quantum-enhanced AI can revolutionize healthcare by enabling personalized medicine and accelerating drug discovery.

- **Personalized Medicine:** Quantum computing can analyse genetic data and medical records at unprecedented speeds, helping to identify personalized treatment plans based on individual patient profiles.
  - **Example:** Quantum-enhanced AI can predict patient responses to different treatments, optimizing therapy plans for conditions like cancer and chronic diseases.

- **Accelerated Drug Discovery:** Quantum computing can simulate molecular interactions and optimize drug candidates more efficiently than classical methods, reducing the time and cost of drug development.
  - **Example:** Companies like IBM and Google are exploring quantum computing for drug discovery,

aiming to find new treatments for diseases such as Alzheimer's and Parkinson's.

Finance and Risk Management: Quantum computing can transform finance by improving financial modeling, risk assessment, and fraud detection.

- **Financial Modeling:** Quantum-enhanced AI can optimize portfolio management, pricing of complex derivatives, and market predictions by processing vast amounts of financial data.
    - **Example:** Quantum algorithms can analyse market trends and provide more accurate predictions for investment strategies, enhancing returns and reducing risks.

- **Risk Assessment:** Quantum computing can enhance risk assessment models by processing complex datasets and identifying hidden correlations, improving the accuracy of credit scoring, and risk management.
    - **Example:** Financial institutions can use quantum-enhanced AI to assess credit risk more accurately, reducing defaults and improving loan approval processes.

- **Fraud Detection:** Quantum-enhanced AI can analyse transaction data in real-time, detecting anomalies and fraudulent activities more effectively.
    - **Example:** Quantum algorithms can identify patterns of fraudulent behavior in large datasets, helping financial institutions prevent and mitigate fraud.

Conclusion: The convergence of quantum computing and AI holds immense potential for transforming various industries by enhancing speed, efficiency, and problem-solving capabilities. Quantum-enhanced AI can tackle complex optimization problems, perform large-scale data analysis, and advance neural networks, leading to significant breakthroughs in healthcare, finance, and beyond. As quantum computing technology continues to evolve, its integration with AI will unlock new possibilities, reshaping our future in profound ways.

This comprehensive exploration of how quantum computing will enhance AI capabilities highlights the transformative potential of combining these two cutting-edge technologies. By examining specific examples and applications, readers can appreciate the profound impact of quantum-enhanced AI on various industries and the innovative solutions it can bring to complex problems.

4.2 Role of Blockchain in Securing AI-Driven Systems

As artificial intelligence (AI) systems become increasingly integrated into various industries, ensuring their security and integrity becomes paramount. Blockchain technology offers robust solutions to address security challenges in AI-driven systems. This section explores the role of blockchain in securing AI, focusing on vulnerabilities in AI systems, blockchain's capability to ensure data integrity, decentralized AI marketplaces, enhancing privacy, and real-world use cases.

4.2.1 Security Challenges in AI

Vulnerabilities in AI Systems: AI systems are susceptible to several security vulnerabilities that can undermine their reliability and effectiveness.

- **Data Poisoning:** In data poisoning attacks, malicious actors introduce corrupt or misleading data into the training datasets used to develop AI models. This can cause the AI to make incorrect decisions or predictions.

- **Example:** In a data poisoning attack on an autonomous vehicle's AI, incorrect data about road conditions could lead to dangerous driving behaviors.

- **Adversarial Attacks:** Adversarial attacks involve subtly modifying inputs to an AI system to cause it to make errors. These attacks exploit the weaknesses in AI algorithms, making them see or interpret things incorrectly.
  - **Example:** Adding noise to an image that is imperceptible to humans but causes an AI system to misclassify it, such as making a stop sign appear as a yield sign to an autonomous vehicle's vision system.

Need for Trustworthy AI: Ensuring the integrity and trustworthiness of AI models and their outputs is crucial for their widespread adoption and reliability.

- **Model Integrity:** Ensuring that AI models have not been tampered with is essential for maintaining their accuracy and reliability.
- **Transparent Decision-Making:** Trustworthy AI systems should provide transparency in their decision-making processes, allowing users to understand and verify how decisions are made.
- **Accountability:** Establishing accountability mechanisms for AI systems helps ensure that any issues or failures can be traced and addressed appropriately.

4.2.2 Blockchain for Data Integrity

Immutable Ledgers: Blockchain's immutable ledger technology ensures that data recorded on the blockchain cannot be altered or deleted, providing a reliable method for ensuring data integrity.

- **Data Provenance:** Blockchain can track the origin and history of data used in AI training, ensuring that only verified and untampered data is used.

- **Example:** In healthcare, blockchain can ensure the integrity of patient data used to train AI models for diagnostics, providing a tamper-proof record of data provenance.

Secure Data Sharing: Blockchain enables secure and transparent data sharing across multiple stakeholders, facilitating collaboration while maintaining data integrity.

- **Distributed Ledger:** By maintaining a distributed ledger, blockchain allows all parties to have access to the same, consistent data, reducing the risk of discrepancies and data tampering.
- **Access Control:** Blockchain can enforce strict access controls, ensuring that only authorized parties can access or modify data.
  - **Example:** In supply chain management, blockchain can securely share data about product provenance and movement, ensuring that all stakeholders have accurate and consistent information.

4.2.3 Decentralized AI Marketplaces

Tokenized Data Economy: Decentralized AI marketplaces leverage blockchain technology to enable the secure exchange of data and AI models using blockchain tokens.

- **Data Monetization:** Data providers can tokenize their data, allowing them to monetize it while maintaining control over how it is used.
- **Model Exchange:** AI developers can securely share and sell their models on decentralized marketplaces, ensuring that transactions are transparent and verifiable.
  - **Example:** Ocean Protocol is a decentralized data exchange protocol that allows data owners to share and monetize their data while retaining control over its usage.

Smart Contracts for AI Services: Smart contracts can automate and enforce agreements for AI services, ensuring transparency and reducing the risk of fraud.

- **Automated Agreements:** Smart contracts can automatically execute transactions when predefined conditions are met, ensuring that agreements between data providers, AI developers, and end-users are honored.
- **Fraud Prevention:** By providing a transparent and tamper-proof record of transactions, smart contracts reduce the risk of fraud and disputes.
  - **Example:** SingularityNET uses smart contracts to facilitate the exchange of AI services on its decentralized platform, ensuring that service agreements are executed automatically and transparently.

4.2.4 Enhancing Privacy with Blockchain

Privacy-Preserving Techniques: Integrating privacy-preserving techniques with blockchain can protect sensitive data while enabling its use in AI systems.

- **Differential Privacy:** Differential privacy techniques add noise to data, ensuring that individual data points cannot be re-identified while still allowing useful insights to be derived.
  - **Example:** Differential privacy can be used to anonymize patient data in healthcare applications, enabling AI models to analyse data without compromising patient privacy.

- **Homomorphic Encryption:** Homomorphic encryption allows computations to be performed on encrypted data without decrypting it, ensuring data privacy throughout the process.
  - **Example:** Homomorphic encryption can enable secure data analysis in financial services, allowing

sensitive financial data to be processed without exposing it to unauthorized parties.

Decentralized Identity Management: Blockchain can provide secure and decentralized identity management systems, ensuring that identities are verified and protected in AI systems.

- **Self-Sovereign Identity:** Decentralized identity solutions give individuals control over their personal information, allowing them to manage and share their identity attributes securely.
    - **Example:** Sovrin is a decentralized identity network that uses blockchain to create self-sovereign identities, allowing users to control their identity data and share it securely.

4.2.5 Case Studies and Use Cases

Healthcare Data Management: Blockchain can secure patient data used in AI-driven healthcare applications, ensuring data integrity and privacy.

- **Example: Medicalchain:** Medicalchain uses blockchain to create secure, patient-centric health records that can be shared with healthcare providers while maintaining patient privacy. AI-driven applications can analyse this data to provide personalized treatment plans and improve patient outcomes.

Supply Chain Automation: Combining blockchain and AI can create more secure and efficient supply chains, enhancing transparency and trust.

- **Example: IBM Food Trust:** IBM Food Trust uses blockchain to track food products from farm to table, ensuring transparency and traceability. AI algorithms analyse this data to optimize supply chain processes, reduce waste, and improve food safety.

- **Example: Maersk's TradeLens:** TradeLens integrates blockchain and AI to streamline global supply chain operations. Blockchain ensures data integrity and transparency, while AI analyses shipping data to optimize routes and improve efficiency.

Conclusion: Blockchain technology provides robust solutions to enhance the security and integrity of AI-driven systems. By leveraging immutable ledgers, secure data sharing, decentralized marketplaces, privacy-preserving techniques, and decentralized identity management, blockchain can address the security challenges inherent in AI systems. Real-world use cases in healthcare, supply chain automation, and other industries demonstrate the transformative potential of integrating blockchain and AI. As these technologies continue to evolve, their combined impact will drive innovation and create more secure, efficient, and trustworthy systems.

This comprehensive exploration of the role of blockchain in securing AI-driven systems highlights its transformative potential in addressing security challenges and enhancing data integrity, privacy, and trust. By examining specific examples and use cases, readers can appreciate how blockchain technology is reshaping the landscape of AI security and its applications across various industries.

4.3 Interoperability and Integration of Quantum, AI, and Blockchain Technologies

The convergence of quantum computing, artificial intelligence (AI), and blockchain technologies holds immense potential for transforming industries and everyday life. However, achieving seamless interoperability and integration among these technologies presents significant challenges. This section explores the challenges of interoperability, the development of hybrid architectures, cross-disciplinary research initiatives, and future directions in this evolving technological ecosystem.

4.3.1 Interoperability Challenges

Technological Silos: One of the primary challenges in integrating quantum computing, AI, and blockchain technologies is their distinct architectures and protocols. Each technology operates within its own silo, making interoperability complex.

- **Distinct Architectures:** Quantum computers use qubits and quantum gates, AI systems leverage neural networks and machine learning algorithms, and blockchain relies on decentralized ledgers and cryptographic protocols. These differing architectures require sophisticated integration methods.
- **Protocol Incompatibility:** The protocols and data formats used by these technologies are often incompatible. For example, quantum computing's need for specialized hardware and algorithms contrasts with the more generalized computing requirements of AI and blockchain systems.
- **Resource Management:** Efficiently managing the computational resources and data flows between quantum computers, AI systems, and blockchain networks is another significant challenge.

Standardization Efforts: To overcome these challenges, there are ongoing efforts to standardize protocols and interfaces for better interoperability.

- **Open Standards:** Developing open standards for data formats, communication protocols, and APIs can facilitate interoperability. Organizations and consortia are working on establishing these standards to enable seamless integration.
    - **Example:** The Institute of Electrical and Electronics Engineers (IEEE) and the International Organization for Standardization (ISO) are involved in creating standards for AI and blockchain technologies.

- **Interoperability Frameworks:** Frameworks that define common interfaces and protocols for data exchange and system integration are crucial. These frameworks ensure that different technologies can work together harmoniously.
  - **Example:** The Hyperledger project, under the Linux Foundation, is developing frameworks to enhance blockchain interoperability with other technologies.

4.3.2 Hybrid Architectures

Quantum-AI Platforms: Developing platforms that seamlessly integrate quantum computing with AI frameworks is essential for harnessing the full potential of these technologies.

- **Quantum Machine Learning (QML):** Quantum-AI platforms facilitate the development of quantum machine learning algorithms that can leverage quantum computing's computational power to enhance AI capabilities.
  - **Example:** IBM's Qiskit framework integrates quantum computing with AI tools, allowing developers to create and test quantum machine learning models.

- **Quantum Enhanced AI Applications:** These platforms enable the deployment of AI applications that benefit from quantum computing's ability to process complex calculations and large datasets more efficiently.
  - **Example:** Google's TensorFlow Quantum integrates quantum computing capabilities with the TensorFlow AI framework, enabling the development of quantum-enhanced AI applications.

Blockchain-Orchestrated Workflows: Blockchain can play a crucial role in orchestrating workflows that involve both quantum computing and AI processes, ensuring secure, transparent, and efficient operations.

- **Smart Contracts for Workflow Automation:** Smart contracts can automate and enforce agreements between different components of the workflow, such as triggering quantum computations based on AI analysis results or recording the outcomes on a blockchain ledger.
  - **Example:** A healthcare application could use blockchain to manage patient consent for AI-driven diagnostics, with quantum computing used for complex genetic analysis.
- **Data Integrity and Provenance:** Blockchain ensures the integrity and provenance of data used in AI and quantum computing, providing a tamper-proof record of all transactions and computations.
  - **Example:** In supply chain management, blockchain can track the provenance of products, AI can optimize logistics, and quantum computing can solve complex optimization problems, all within an integrated workflow.

4.3.3 Cross-Disciplinary Research

Collaborative Research Initiatives: Collaborative research initiatives and consortia are essential for driving innovation at the intersection of quantum computing, AI, and blockchain.

- **Interdisciplinary Collaboration:** Researchers from different fields work together to explore new applications and solve complex problems that span multiple disciplines.
  - **Example:** The Quantum AI Lab (QuAIL) at NASA Ames Research Center collaborates with academic institutions and technology companies to explore quantum computing's applications in AI.
- **Industry-Academia Partnerships:** Partnerships between industry and academia foster the development of cutting-edge technologies and accelerate their commercialization.

- **Example:** The IBM Q Network is a collaborative network of Fortune 500 companies, academic institutions, and national research labs working together to advance quantum computing and its integration with AI and blockchain.

Funding and Investment: Funding and investment trends are crucial for supporting cross-disciplinary research and development in quantum computing, AI, and blockchain.

- **Government Grants:** Governments are providing grants and funding for research projects that explore the convergence of these technologies.
    - **Example:** The U.S. National Science Foundation (NSF) offers grants for research in quantum computing, AI, and blockchain, encouraging interdisciplinary projects.
- **Venture Capital Investment:** Venture capital firms are increasingly investing in startups and companies developing innovative solutions at the intersection of these technologies.
    - **Example:** Quantum computing startups like Rigetti Computing and D-Wave Systems have attracted significant investment to develop quantum hardware and integrate it with AI and blockchain solutions.

4.3.4 Future Directions

Evolving Ecosystem: The technological ecosystem where quantum computing, AI, and blockchain are tightly integrated is rapidly evolving, with exciting prospects on the horizon.

- **Convergence of Technologies:** The convergence of these technologies will lead to the creation of new platforms, applications, and services that leverage the strengths of each technology.

- **Example:** Future healthcare systems could use quantum computing for complex genetic analysis, AI for personalized treatment plans, and blockchain for secure data sharing and patient consent management.

- **New Business Models:** The integration of these technologies will enable new business models that offer innovative solutions to complex problems.

    - **Example:** Decentralized AI marketplaces powered by blockchain could allow data owners, AI developers, and quantum computing providers to collaborate and create value in a transparent and secure environment.

Potential Breakthroughs: The convergence of quantum computing, AI, and blockchain holds the promise of breakthrough innovations and game-changing advancements.

- **Quantum-Enhanced AI:** Quantum computing could enable AI to solve problems that are currently intractable, leading to breakthroughs in fields such as drug discovery, climate modeling, and financial optimization.

    - **Example:** Quantum-enhanced AI could develop new materials with specific properties for use in industries like aerospace, energy, and healthcare.

- **Secure and Transparent Systems:** Blockchain's ability to provide secure and transparent records could revolutionize sectors like finance, supply chain, and governance.

    - **Example:** A decentralized, blockchain-based voting system powered by quantum-enhanced AI could ensure secure, transparent, and tamper-proof elections.

- **Interdisciplinary Innovations:** The fusion of these technologies will drive interdisciplinary innovations, leading to the development of new tools, frameworks, and methodologies.

- **Example:** New programming languages and development environments tailored for quantum-AI applications could emerge, simplifying the creation of complex, integrated systems.

Conclusion: The integration of quantum computing, AI, and blockchain technologies presents both significant challenges and immense opportunities. Overcoming interoperability challenges, developing hybrid architectures, fostering cross-disciplinary research, and supporting funding and investment are crucial for driving innovation in this evolving technological ecosystem. The convergence of these technologies promises to unlock new possibilities, leading to breakthroughs that will reshape industries and everyday life over the next 10 to 20 years.

This comprehensive exploration of the interoperability and integration of quantum computing, AI, and blockchain technologies highlights the challenges and future directions in this rapidly evolving field. By examining specific examples and potential breakthroughs, readers can appreciate the transformative potential of these technologies and the innovative solutions they can bring to complex problems.

4.4 Ethical and Societal Implications of Technology Convergence

The convergence of quantum computing, artificial intelligence (AI), and blockchain technology presents profound ethical and societal implications. As these technologies reshape various industries and everyday life, it is crucial to address ethical considerations, societal impacts, regulatory frameworks, and preparations for the future to ensure they benefit all of humanity. This section explores these dimensions in detail.

4.4.1 Ethical Considerations

Bias and Fairness: AI systems, especially those enhanced by quantum computing and integrated with blockchain, must be designed to mitigate biases and ensure fairness.

- **AI Model Bias:** AI models can inadvertently perpetuate biases present in training data, leading to unfair or discriminatory outcomes. It is essential to implement robust methods for identifying and correcting biases in AI models.

    - **Example:** In hiring algorithms, biases can lead to discriminatory hiring practices if the training data reflects historical biases. Techniques like fairness-aware machine learning can help mitigate these issues.

- **Fairness in Quantum and Blockchain Technologies:** Ensuring fairness extends to quantum computing and blockchain as well. Quantum algorithms should be designed to provide equitable outcomes, and blockchain systems must ensure fair access and participation.

    - **Example:** Fairness in blockchain involves preventing unfair advantages in consensus mechanisms (e.g., preventing the concentration of mining power in proof-of-work systems).

Transparency and Accountability: Maintaining transparency and accountability is critical in the development and deployment of converged technologies.

- **Algorithmic Transparency:** AI systems, especially those using quantum-enhanced algorithms, should be transparent about how decisions are made. This includes providing explanations for AI-driven decisions to foster trust and accountability.

    - **Example:** Implementing explainable AI (XAI) techniques can help users understand the reasoning behind AI decisions, enhancing transparency.

- **Blockchain Accountability:** Blockchain's immutable ledger can enhance accountability by providing a transparent and tamper-proof record of transactions and decisions. This is particularly important for ensuring the integrity of AI training data and decision-making processes.

    - **Example:** In supply chain management, blockchain can provide an auditable trail of all transactions, ensuring accountability at every stage.

4.4.2 Impact on Society

Job Displacement and Creation: The convergence of advanced technologies will significantly impact the job market, leading to both job displacement and the creation of new job opportunities.

- **Job Displacement:** Automation driven by AI and quantum computing may displace jobs, particularly in industries reliant on routine, manual tasks. Workers in these sectors may face challenges as their roles become automated.
  - **Example:** Manufacturing jobs may decline as AI-driven robots and quantum-optimized production processes become more prevalent.
- **Job Creation:** Conversely, new jobs will emerge in fields such as quantum computing, AI development, blockchain engineering, and cybersecurity. These roles will require advanced technical skills and expertise.
  - **Example:** Roles such as quantum algorithm developers, blockchain architects, and AI ethicists will become increasingly important.

Digital Divide: Addressing the digital divide is crucial to ensure equitable access to advanced technologies.

- **Access to Technology:** Disparities in access to technology can exacerbate existing inequalities. Efforts must be made to ensure that all communities have access to the benefits of quantum computing, AI, and blockchain.
  - **Example:** Programs that provide access to high-speed internet, affordable computing devices, and digital literacy training can help bridge the digital divide.

- **Inclusive Development:** Inclusive development practices should be adopted to ensure that the development and deployment of converged technologies benefit all segments of society.

  - **Example:** Community-driven technology initiatives that involve local stakeholders in decision-making processes can promote inclusivity.

4.4.3 Regulatory and Policy Frameworks

Governance Models: Effective governance models are needed to oversee the ethical and responsible use of converged technologies.

- **Ethical Oversight:** Regulatory bodies and ethical committees should be established to monitor the development and deployment of advanced technologies, ensuring they adhere to ethical standards.

  - **Example:** An independent ethics board could review AI applications to ensure they do not perpetuate biases or harm societal interests.

- **Flexible Regulations:** Regulations should be flexible enough to adapt to the rapid pace of technological advancements while ensuring that fundamental ethical principles are upheld.

  - **Example:** Regulatory sandboxes can provide a controlled environment for testing new technologies and business models under regulatory supervision.

International Collaboration: International collaboration is essential for developing coherent regulatory frameworks and standards for converged technologies.

- **Global Standards:** Developing global standards for quantum computing, AI, and blockchain can facilitate interoperability, enhance security, and ensure consistent ethical practices.
    - **Example:** The International Telecommunication Union (ITU) works on global standards for emerging technologies, promoting international collaboration.
- **Cross-Border Collaboration:** Collaboration between countries can address transnational challenges, such as cybersecurity threats and data privacy concerns, that arise from the deployment of advanced technologies.
    - **Example:** International agreements on data protection and cybersecurity can enhance global security and trust in converged technologies.

4.4.4 Preparing for the Future

Education and Workforce Development: Preparing individuals for careers in emerging technology fields is crucial for harnessing the potential of quantum computing, AI, and blockchain.

- **Curriculum Development:** Educational institutions should develop curricula that cover the fundamentals of these technologies and their applications, fostering a skilled workforce.

- **Example:** Universities offering specialized programs in quantum computing, AI, and blockchain can equip students with the necessary knowledge and skills.

- **Lifelong Learning:** Lifelong learning programs and professional development opportunities should be available to help workers transition to new roles created by technology convergence.

  - **Example:** Online courses and certification programs in AI, blockchain, and quantum computing can help professionals upskill and stay competitive in the job market.

Public Awareness and Engagement: Increasing public awareness and engagement with advanced technologies and their implications is essential for informed decision-making and societal acceptance.

- **Public Education Campaigns:** Public education campaigns can raise awareness about the benefits and challenges of converged technologies, promoting informed public discourse.

  - **Example:** Government and non-profit organizations can organize workshops, webinars, and public talks to educate the public about emerging technologies.

- **Engagement Platforms:** Creating platforms for public engagement can facilitate dialogue between technologists, policymakers, and the public, ensuring that diverse perspectives are considered in technology development.

- **Example:** Online forums and community meetings can provide opportunities for stakeholders to discuss the ethical and societal implications of advanced technologies.

Conclusion: The convergence of quantum computing, AI, and blockchain technologies presents both opportunities and challenges. Addressing ethical considerations, understanding the societal impact, developing robust regulatory frameworks, and preparing for the future through education and public engagement are crucial for ensuring that these technologies benefit all of humanity. By proactively addressing these issues, we can harness the potential of these cutting-edge technologies to create a more equitable, secure, and prosperous future.

This comprehensive exploration of the ethical and societal implications of technology convergence highlights the importance of addressing biases, ensuring transparency, preparing for job market changes, bridging the digital divide, developing effective regulatory frameworks, and engaging the public. By examining specific examples and strategies, readers can appreciate the critical role these considerations play in shaping the future of quantum computing, AI, and blockchain technologies.

Chapter 5: Applications and Industry Impacts

5.1 Healthcare

The healthcare industry stands to gain significantly from the convergence of quantum computing, artificial intelligence (AI), and blockchain technology. These advanced technologies offer transformative solutions that can revolutionize medical diagnosis and treatment, improve health data management, and accelerate drug discovery. This section delves into the profound impact these technologies are having on healthcare, illustrated with case studies and real-world examples.

5.1.1 Revolutionizing Medical Diagnosis and Treatment

AI-Driven Diagnostics: AI algorithms are making significant strides in medical diagnostics, providing tools for early detection of diseases such as cancer and heart disease through image analysis and predictive analytics.

- **Case Study: AI in Radiology**
    - **Early Detection of Cancer:** AI algorithms are being trained to analyse medical images, such as mammograms and CT scans, to detect early signs of cancer. For instance, Google's DeepMind

developed an AI system that can outperform human radiologists in identifying breast cancer from mammograms.

- **Heart Disease Prediction:** AI-driven predictive analytics can analyse patient data, including electronic health records (EHRs) and genetic information, to predict the risk of heart disease. IBM Watson Health has developed AI tools that assist cardiologists in identifying high-risk patients and recommending personalized treatment plans.

Personalized Medicine: AI and quantum computing are playing pivotal roles in advancing personalized medicine by analysing genetic information and tailoring treatments to individual patients.

- **Real-World Example: Precision Oncology**
    - **Genetic Profiling:** AI algorithms can analyse a patient's genetic profile to identify specific mutations and biomarkers associated with cancer. Companies like Tempus are leveraging AI to provide oncologists with insights into the most effective treatments based on a patient's unique genetic makeup.
    - **Quantum Computing for Personalized Treatment:** Quantum computers can process vast amounts of genetic data much faster than classical computers, identifying optimal treatment pathways. For example, quantum algorithms can simulate how different drugs interact with a patient's unique genetic

mutations, enabling highly personalized treatment plans.

5.1.2 Blockchain for Health Data Management

Secure Patient Records: Blockchain technology offers robust solutions for securing patient records, ensuring privacy, and enabling efficient data sharing among healthcare providers.

- **Case Study: MedRec**
    - **MedRec Platform:** Developed by researchers at MIT, MedRec uses blockchain to create a secure and immutable ledger of patient records. This platform allows patients to have control over their data while enabling healthcare providers to access accurate and up-to-date information. MedRec ensures data integrity, reduces administrative costs, and enhances patient privacy.

Drug Supply Chain Integrity: Blockchain can enhance the integrity of the pharmaceutical supply chain by tracking medications from production to delivery, preventing counterfeit drugs, and ensuring authenticity.

- **Use Case: PharmaLedger**
    - **PharmaLedger Project:** This blockchain consortium includes major pharmaceutical companies and aims to improve the traceability of drugs through the supply chain. By recording every transaction on a blockchain, PharmaLedger ensures that each medication is authentic and safe for consumption. This reduces the risk of counterfeit drugs entering the market and

ensures that patients receive genuine medications.

5.1.3 Quantum Computing in Drug Discovery

Accelerated Research: Quantum computing has the potential to significantly speed up the process of drug discovery by simulating molecular interactions more efficiently than classical computers.

- **Example: Quantum Simulations**
    - **Molecular Modeling:** Quantum computers can simulate the behavior of molecules at the quantum level, providing insights into their interactions that are beyond the capabilities of classical computers. This accelerates the identification of potential drug candidates and reduces the time required for preclinical testing.

Case Study: Quantum Drug Discovery

- **Pharmaceutical Company: Biogen**
    - **Quantum Research Collaboration:** Biogen has partnered with quantum computing companies like 1QBit to explore the use of quantum computing in drug discovery. By leveraging quantum algorithms, Biogen aims to develop new treatments for complex diseases such as Alzheimer's. The quantum simulations help in understanding the misfolding of proteins associated with the disease, paving the way for novel therapeutic approaches.

Conclusion: The convergence of quantum computing, AI, and blockchain is poised to revolutionize healthcare by enhancing medical diagnostics, personalizing treatments, securing health data, and accelerating drug discovery. Through case studies and real-world applications, this section illustrates how these cutting-edge technologies are transforming healthcare, offering new hope for improved patient outcomes and more efficient medical practices.

This comprehensive exploration of healthcare highlights the transformative potential of integrating quantum computing, AI, and blockchain technologies. By examining specific examples and case studies, readers can appreciate the profound impact these technologies are having on medical diagnosis, treatment, data management, and drug discovery.

5.2 Finance

The financial industry is undergoing a profound transformation driven by the convergence of quantum computing, artificial intelligence (AI), and blockchain technology. These advanced technologies offer innovative solutions for algorithmic trading, risk management, decentralized finance (DeFi), cross-border payments, financial modeling, and fraud detection. This section explores how these technologies are reshaping the financial landscape, illustrated with case studies and real-world examples.

5.2.1 AI in Financial Services

Algorithmic Trading: AI and machine learning algorithms are revolutionizing trading strategies, enabling faster and more accurate decision-making.

- **Case Study: Renaissance Technologies**
    - **Medallion Fund:** Renaissance Technologies, a hedge fund known for its quantitative trading strategies, uses AI and machine learning algorithms to analyse vast amounts of market data. The Medallion Fund, one of its most successful funds, employs sophisticated algorithms to identify trading opportunities and execute trades at high speed and precision.
    - **Impact:** AI-driven algorithmic trading enables Renaissance Technologies to consistently outperform traditional trading strategies, achieving remarkable returns even in volatile market conditions.

Risk Management: AI is transforming risk management by assessing and mitigating financial risks through the analysis of large datasets and prediction of market trends.

- **Example: BlackRock's Aladdin**
    - **Aladdin Platform:** BlackRock's Aladdin (Asset, Liability, Debt, and Derivative Investment Network) is an AI-powered risk management system used by financial institutions worldwide. Aladdin analyses extensive datasets to identify potential risks, simulate market scenarios, and provide insights for better decision-making.
    - **Impact:** The platform enhances risk management by enabling financial institutions to proactively address risks, optimize asset allocations, and improve investment outcomes.

5.2.2 Blockchain for Financial Transactions

Decentralized Finance (DeFi): DeFi platforms use blockchain to offer financial services such as lending, borrowing, and trading without traditional intermediaries.

- **Real-World Example: Aave**

- **Aave Platform:** Aave is a decentralized lending platform built on the Ethereum blockchain. It allows users to lend and borrow cryptocurrencies without relying on traditional banks. Users can earn interest on their deposits and borrow assets by providing collateral.
- **Impact:** Aave democratizes access to financial services, reduces costs, and increases transparency by eliminating intermediaries. The platform has grown significantly, managing billions of dollars in assets.

Cross-Border Payments: Blockchain technology facilitates faster and cheaper cross-border transactions by reducing reliance on intermediaries and streamlining processes.

- **Case Study: Ripple and Santander**
    - **RippleNet:** Ripple's blockchain-based payment network, RippleNet, enables real-time cross-border transactions. Santander, a global bank, uses RippleNet's technology for its One Pay FX service, which allows customers to make international payments quickly and cost-effectively.
    - **Impact:** RippleNet significantly reduces transaction times and costs compared to traditional banking methods, enhancing the efficiency of cross-border payments and providing a better customer experience.

5.2.3 Quantum Computing for Financial Modeling

Portfolio Optimization: Quantum computing can optimize investment portfolios by efficiently solving complex mathematical problems that are computationally intensive for classical computers.

- **Example: Portfolio Optimization Algorithms**
    - **Quantum Algorithms:** Quantum algorithms, such as the Quantum Approximate Optimization Algorithm (QAOA), can handle complex portfolio optimization

> problems by exploring numerous combinations of assets to identify the optimal portfolio that maximizes returns and minimizes risks.
>
> - **Impact:** Quantum computing enables financial institutions to perform more accurate and efficient portfolio optimization, leading to better investment strategies and improved financial performance.

Fraud Detection: Quantum-enhanced algorithms improve the detection of fraudulent activities in financial transactions by analysing vast amounts of data and identifying patterns indicative of fraud.

- **Example: Quantum-Enhanced Machine Learning**
    - **Fraud Detection Models:** Financial institutions can use quantum-enhanced machine learning models to detect fraudulent transactions by identifying subtle correlations and patterns that classical algorithms might miss. These models can analyse transaction data in real-time, providing more accurate and timely fraud detection.
    - **Impact:** Enhanced fraud detection capabilities help financial institutions reduce losses due to fraud, protect customer assets, and maintain trust in their services.

Conclusion: The convergence of quantum computing, AI, and blockchain is transforming the financial industry by enhancing trading strategies, risk management, financial transactions, portfolio optimization, and fraud detection. Through case studies and real-world examples, this section illustrates how these cutting-edge technologies are reshaping finance, offering innovative solutions that improve efficiency, reduce costs, and enhance security. As these technologies continue to evolve, their impact on the financial industry will only grow, driving further advancements and opportunities.

This comprehensive exploration of finance highlights the transformative potential of integrating quantum computing, AI, and blockchain technologies. By examining specific examples and case studies, readers can appreciate the profound impact these technologies are having on trading, risk management, financial transactions, and fraud detection.

5.3 Urban Planning

Urban planning is a critical aspect of modern society, shaping the environments where people live, work, and interact. The integration of quantum computing, artificial intelligence (AI), and blockchain technology offers transformative solutions for urban planning, enabling smarter, more efficient, and more transparent cities. This section explores how these technologies are revolutionizing urban planning, with detailed case studies and practical examples.

5.3.1 AI and Smart Cities

Traffic Management: AI is being employed to analyse traffic patterns and optimize traffic flow, reducing congestion and emissions in urban areas.

- **Case Study: Barcelona's Smart Traffic Management System**
    - **Implementation:** Barcelona has implemented an AI-driven traffic management system that uses data from cameras, sensors, and GPS devices to monitor traffic in real-time. The system analyses traffic patterns and predicts congestion, adjusting traffic signals and rerouting traffic to optimize flow.
    - **Impact:** This AI system has significantly reduced traffic congestion and emissions, improving air quality and reducing commute times for residents. It has also enhanced the efficiency of public

transportation, making it a more attractive option for commuters.

Resource Allocation: AI is also being used to manage urban resources more efficiently, optimizing the use of energy, water, and waste management.

- **Example: Singapore's Smart Nation Initiative**
    - **Energy Management:** AI systems monitor energy consumption across the city, optimizing the operation of heating, ventilation, and air conditioning (HVAC) systems in buildings to reduce energy use and costs. Predictive analytics are used to forecast energy demand and adjust supply accordingly.
    - **Water Management:** AI technologies manage water distribution networks, detecting leaks and predicting maintenance needs to ensure a reliable supply of clean water. Machine learning algorithms analyse weather data and water usage patterns to optimize irrigation schedules for public parks and gardens.
    - **Waste Management:** AI-driven waste management systems use sensors to monitor waste levels in bins across the city, optimizing collection routes and schedules to improve efficiency and reduce fuel consumption.

5.3.2 Blockchain for Urban Development

Transparent Land Registry: Blockchain technology is being used to create transparent and tamper-proof land registry systems, ensuring accurate and secure property records.

- **Example: Sweden's Blockchain Land Registry**
    - **Implementation:** Sweden's Lantmäteriet (Land Registry) has developed a blockchain-based system to record property transactions. The system creates an immutable record of ownership changes, ensuring transparency and security.

- **Impact:** This blockchain system reduces fraud and errors in property transactions, streamlines the process of buying and selling property, and increases trust among stakeholders. It also speeds up the registration process and reduces administrative costs.

Public Participation: Blockchain is also facilitating public participation in urban planning decisions, enhancing transparency and accountability.

- **Example: Decidim in Barcelona**
    - **Implementation:** Decidim is a blockchain-based platform used by the city of Barcelona to engage citizens in the urban planning process. The platform allows residents to propose, debate, and vote on urban development projects.
    - **Impact:** Decidim ensures that public participation is transparent and tamper-proof, increasing trust in the decision-making process. It empowers citizens to have a direct say in the development of their city, fostering a sense of community and collaboration.

5.3.3 Quantum Computing for Urban Simulation

Complex Simulations: Quantum computing can simulate urban environments and predict the impact of various planning decisions more accurately than classical methods.

- **Capabilities:** Quantum computers can process and analyse vast amounts of data simultaneously, making them ideal for complex simulations of urban environments. They can model the interactions between different variables, such as traffic flow, energy consumption, and population growth, to predict the outcomes of various planning scenarios.
    - **Example:** A city planning to build a new transportation network can use quantum simulations to assess the impact on traffic congestion, pollution

levels, and economic activity, ensuring that the project is optimized for maximum benefit.

Case Study: Quantum Urban Planning in New York City

- **Implementation:** New York City is exploring the use of quantum computing to optimize its infrastructure projects. By partnering with quantum computing firms, the city uses quantum simulations to model the impact of new developments on traffic patterns, energy usage, and environmental sustainability.
  - **Impact:** Quantum simulations have enabled New York City to make more informed decisions about infrastructure projects, reducing costs and minimizing negative impacts. The simulations have also helped identify the most effective strategies for enhancing resilience to climate change and improving urban livability.

Conclusion: The convergence of quantum computing, AI, and blockchain technology is revolutionizing urban planning, enabling smarter, more efficient, and more transparent cities. AI-driven systems are optimizing traffic management and resource allocation, blockchain is ensuring transparent land registries and public participation, and quantum computing is providing accurate simulations for better planning decisions. Through detailed case studies and practical examples, this section illustrates how these technologies are transforming urban environments and shaping the cities of the future.

This comprehensive exploration of urban planning highlights the transformative potential of integrating quantum computing, AI, and blockchain technologies. By examining specific examples and case studies, readers can appreciate the profound impact these technologies are having on traffic management, resource allocation, land registries, public participation, and urban simulations.

- .

5.4 Public Policy

The intersection of quantum computing, artificial intelligence (AI), and blockchain technology, coupled with the emerging concept of the Metaverse of the Minds Social Direct Democracy (MOTMSDD), promises to revolutionize public policy. These advanced technologies, together with MOTMSDD, offer powerful tools for predictive analytics, transparent governance, complex policy modeling, and direct public engagement. This section explores how these technologies can enhance public policy through detailed case studies and practical examples, while integrating the future use of MOTMSDD.

5.4.1 AI for Policy Analysis

Predictive Analytics: AI's predictive analytics capabilities are transforming how policymakers evaluate the potential outcomes of different policy decisions, enabling more informed and effective governance.

- **Case Study: Predictive Analytics in Healthcare Policy**
    - **Implementation:** The UK National Health Service (NHS) uses AI-driven predictive analytics to forecast the impact of various healthcare policies. By analysing historical data and current trends, AI models predict outcomes such as hospital admission rates, patient waiting times, and overall healthcare costs.
    - **Impact:** These predictions help policymakers make evidence-based decisions, optimize resource allocation, and improve patient care. For example, predictive analytics can inform the expansion of

vaccination programs or the allocation of funding to under-resourced areas.

Sentiment Analysis: AI is also being used to analyse public sentiment on social media and other platforms, providing insights into public opinion on various issues.

- **Example: AI Sentiment Analysis for Urban Planning**
    - **Implementation:** The city of Boston uses AI sentiment analysis to gauge public opinion on urban planning initiatives. By analysing comments and reactions on social media platforms, AI tools provide real-time insights into citizens' views on new projects, zoning changes, and infrastructure developments.
    - **Impact:** Understanding public sentiment helps policymakers address concerns, build consensus, and tailor policies to better meet the needs and preferences of the community. This approach enhances public trust and participation in the policymaking process.

5.4.2 Blockchain for Transparent Governance

Anti-Corruption Measures: Blockchain technology enhances transparency and reduces corruption by providing immutable records of transactions and processes.

- **Example: Blockchain in Public Procurement in Georgia**
    - **Implementation:** The government of Georgia has implemented a blockchain-based system to manage public procurement processes. This system records all transactions related to government contracts on a blockchain, making them transparent and tamper-proof.
    - **Impact:** The blockchain system has significantly reduced opportunities for corruption and fraud in public procurement, ensuring that government contracts are awarded fairly and transparently. It also

increases public trust in the integrity of government operations.

Digital Voting Systems: Blockchain-based voting systems ensure secure and transparent elections, protecting the integrity of the democratic process.

- **Case Study: Voatz in West Virginia**
    - **Implementation:** In the 2018 midterm elections, West Virginia piloted a blockchain-based mobile voting system called Voatz for overseas military personnel. The system used blockchain to securely record votes and ensure their immutability.
    - **Impact:** The pilot was successful in providing a secure and accessible voting option for military personnel, with the blockchain ensuring the transparency and integrity of the votes. This initiative demonstrates the potential for blockchain to enhance the security and accessibility of elections worldwide.

5.4.3 Quantum Computing for Complex Policy Modeling

Scenario Analysis: Quantum computing can model complex policy scenarios by taking into account numerous variables and their interactions, providing policymakers with deeper insights into potential outcomes.

- **Example: Quantum Modeling for Climate Policy**
    - **Implementation:** Researchers at the Canadian Quantum Computing Company, D-Wave, have developed quantum algorithms to model climate change scenarios. These models consider various factors, including greenhouse gas emissions, economic impacts, and technological advancements.
    - **Impact:** Quantum-enhanced scenario analysis helps policymakers understand the long-term implications of different climate policies, such as carbon pricing or renewable energy subsidies. This allows for more

informed decision-making that balances environmental sustainability with economic growth.

Case Study: Quantum Simulations in Environmental Policy in the Netherlands

- **Implementation:** The Dutch government is using quantum computing to simulate the impact of proposed environmental policies on the country's water management systems. Quantum simulations analyse variables such as sea-level rise, precipitation patterns, and infrastructure resilience.
  - **Impact:** These simulations provide high-resolution insights into how various policies could mitigate the effects of climate change on the Netherlands' intricate water management system. The results inform strategic planning and investment in infrastructure to protect against future environmental challenges.

5.4.4 MOTMSDD for Public Engagement

Direct Democracy through MOTMSDD: The Metaverse of the Minds Social Direct Democracy (MOTMSDD) integrates advanced technologies to facilitate real-time, direct public decision-making, enhancing public engagement in policy development.

- **Digital Twin Representation:** In the MOTMSDD metaverse, each individual has a digital twin that represents their data and preferences. This digital twin acts as a delegate, participating in real-time decision-making processes on behalf of the individual.
  - **Impact:** This ensures that public policy reflects the true needs and preferences of the population, increasing the legitimacy and acceptance of policy decisions.

Brain-Computer Interface (BCI) Integration: BCI technology enables seamless interaction between individuals and the

MOTMSDD platform, allowing for intuitive and direct participation in policymaking.

- **Example: Real-Time Public Consultation**
    - **Implementation:** During a public consultation on urban development, citizens use BCIs to interact with the MOTMSDD platform, expressing their opinions and voting on various proposals. The platform aggregates these inputs, providing policymakers with a comprehensive understanding of public sentiment.
    - **Impact:** This approach enhances public engagement, ensuring that a diverse range of voices is heard and considered in the decision-making process.

Conflict Resolution through AI and Quantum Computing: MOTMSDD leverages AI and quantum computing to resolve conflicts between different stakeholders, optimizing policy decisions to address the true needs of the population.

- **Scenario: Urban Planning Dispute**
    - **Implementation:** In a scenario where there is a dispute over a proposed urban development project, AI analyses the preferences and concerns of all stakeholders, while quantum computing simulates various compromise solutions. The optimal solution that balances the needs of all parties is presented for final approval through the MOTMSDD platform.
    - **Impact:** This process ensures that policy decisions are fair, balanced, and reflective of the true needs and preferences of the population, enhancing social harmony and cooperation.

Conclusion: The convergence of quantum computing, AI, blockchain technology, and the innovative MOTMSDD platform is transforming public policy by enabling predictive analytics, enhancing transparency, providing powerful tools for complex scenario modeling, and facilitating direct public engagement. Through detailed case studies and practical examples, this section illustrates how these technologies are enhancing policy analysis, governance, and decision-making processes. As these technologies continue to evolve, their impact on public policy will become increasingly significant, driving more effective and equitable governance.

This comprehensive exploration of public policy highlights the transformative potential of integrating quantum computing, AI, blockchain technologies, and MOTMSDD. By examining specific examples and case studies, readers can appreciate the profound impact these technologies are having on policy analysis, transparent governance, complex scenario modeling, and direct public engagement.

5.5 Real-World Examples and Potential Future Scenarios

The convergence of quantum computing, artificial intelligence (AI), and blockchain technology presents unprecedented opportunities to reshape industries and everyday life. This section explores real-world examples and potential future scenarios, illustrating the transformative potential of these technologies and the Metaverse of the Minds Social Direct Democracy (MOTMSDD) approach.

5.5.1 Interdisciplinary Case Studies

Healthcare and Finance: An integrated system where blockchain ensures secure health data management while AI analyses financial health records to offer personalized financial advice.

- **Case Study: Integrated Health and Financial Wellness System**
    - **Implementation:** A pioneering health-tech firm collaborates with a fintech company to create an

integrated platform. Blockchain secures patient health records, ensuring privacy and enabling seamless sharing among authorized healthcare providers. Simultaneously, AI analyses the financial health records of individuals to provide tailored financial advice.

- **Impact:** Patients receive comprehensive care that considers both their physical and financial well-being. For example, a patient with a chronic illness can receive personalized financial planning advice to manage medical expenses and optimize insurance benefits, improving overall quality of life.

Urban Planning and Public Policy: A city leveraging AI for smart infrastructure management, blockchain for transparent governance, and quantum computing simulations to forecast long-term impacts.

- **Case Study: Smart City Initiative**
    - **Implementation:** A metropolitan area implements a smart city initiative integrating AI, blockchain, and quantum computing. AI manages urban infrastructure, optimizing traffic flow, energy use, and public services. Blockchain ensures transparent and tamper-proof governance, recording transactions and public decisions. Quantum computing simulations model long-term impacts of urban development projects.
    - **Impact:** The city experiences reduced traffic congestion, optimized energy consumption, and enhanced public trust in governance. Quantum simulations enable policymakers to foresee the environmental and social impacts of new infrastructure projects, ensuring sustainable urban development.

5.5.2 Potential Future Scenarios

Ubiquitous AI and Quantum Integration: A vision of a future where AI, the MOTMSDD approach, and quantum computing are seamlessly integrated into everyday life, enhancing everything from healthcare to urban living.

- **Scenario: A Day in a Quantum-Enhanced Smart City**
    - **Morning:** An individual starts their day with an AI-powered personal assistant that synchronizes their schedule, monitors health metrics, and provides personalized recommendations. Blockchain secures their health data, which is accessible only to authorized healthcare providers.
    - **Afternoon:** At work, quantum computing optimizes complex business decisions, from supply chain logistics to financial modeling. The city uses AI to manage traffic and energy, ensuring smooth commutes and sustainable living.
    - **Evening:** The MOTMSDD platform facilitates direct citizen engagement in public policy, allowing residents to vote on community projects and local regulations in real-time, using brain-computer interfaces for seamless interaction.
    - **Impact:** This integration leads to a highly efficient, participatory, and equitable society where technology enhances all aspects of daily life.

Decentralized Autonomous Organizations (DAOs): Blockchain enables the creation of DAOs that operate without traditional hierarchical management, driven by AI and secured by quantum-resistant cryptography.

- **Scenario: The Rise of DAOs in Business and Governance**
    - **Business:** A new startup leverages blockchain to create a DAO, where AI algorithms manage operations such as payroll, logistics, and customer

service. Quantum-resistant cryptography ensures the security of transactions and data integrity.

- **Governance:** Local governments adopt DAOs for community management, where residents can propose, vote on, and implement local policies directly. AI analyses proposals and predicts outcomes, while blockchain ensures transparency and trust.
- **Impact:** DAOs enable more democratic, transparent, and efficient management structures. Businesses and communities benefit from reduced administrative overhead and increased trust in governance processes.

5.5.3 Ethical and Societal Considerations

Balancing Innovation with Ethics: As we advance technologically, it is crucial to balance innovation with ethical considerations, ensuring that these advancements benefit society as a whole.

- **Discussion:** The rapid development of AI, quantum computing, and blockchain technology raises ethical questions about privacy, security, and inequality. For example, the use of AI in decision-making processes must be transparent and free from bias to avoid reinforcing existing inequalities. Quantum computing's potential to break current cryptographic systems necessitates the development of quantum-resistant security measures to protect sensitive data.
  - **Strategy:** Establishing ethical guidelines and regulatory frameworks that prioritize transparency, accountability, and fairness in the development and deployment of these technologies.

Preparing for Technological Disruption: Strategies for individuals, businesses, and governments to prepare for the disruptive impacts of these converging technologies.

- **Individuals:**

- **Education and Skills Development:** Encouraging lifelong learning and upskilling to adapt to new job roles created by technological advancements. Programs focusing on digital literacy, AI, blockchain, and quantum computing are essential.
- **Adaptability:** Cultivating a mindset of adaptability and resilience to navigate the rapid pace of technological change.

- **Businesses:**
    - **Innovation and Investment:** Investing in research and development to harness the potential of emerging technologies. Collaborating with technology partners and adopting flexible business models to stay competitive.
    - **Ethical Practices:** Implementing ethical guidelines for technology use, ensuring fairness, transparency, and accountability in AI and blockchain applications.

- **Governments:**
    - **Policy and Regulation:** Developing policies and regulatory frameworks that promote innovation while protecting public interests. This includes data protection laws, cybersecurity regulations, and ethical standards for AI and quantum computing.
    - **Public Engagement:** Engaging with the public to raise awareness about the benefits and challenges of emerging technologies. Facilitating open dialogue and participation in decision-making processes.

Conclusion: The convergence of quantum computing, AI, and blockchain, along with the MOTMSDD approach, offers transformative potential for various industries and everyday life. Real-world examples and potential future scenarios illustrate how these technologies can enhance healthcare, finance, urban planning, and public policy. Balancing innovation with ethical considerations and preparing for technological disruption are crucial for ensuring that these advancements benefit society as a whole. As we move towards this future, a comprehensive and forward-thinking approach will be essential to harness the full potential of these cutting-edge technologies.

This comprehensive exploration of real-world examples and potential future scenarios highlights the transformative potential of integrating quantum computing, AI, blockchain technologies, and the MOTMSDD approach. By examining specific examples and future visions, readers can appreciate the profound impact these technologies are having and will continue to have on various aspects of life and industry.

5.6 Industry-Specific Deep Dives

The convergence of quantum computing, artificial intelligence (AI), and blockchain technology is reshaping various industries in profound ways. This section delves deeply into specific industries, highlighting how these cutting-edge technologies are being integrated to drive innovation, efficiency, and sustainability.

5.6.1 Healthcare Deep Dive

Telemedicine and Remote Monitoring: AI and blockchain are revolutionizing telemedicine and remote patient monitoring, extending healthcare access to remote and underserved areas.

- **Implementation:**
    - **AI-Powered Diagnostics:** Telemedicine platforms equipped with AI algorithms can analyse patient data, such as vital signs and medical history, to provide preliminary diagnoses and treatment recommendations. For instance, AI-driven chatbots

can triage patients and suggest next steps based on their symptoms.
- **Blockchain for Data Security:** Blockchain ensures the security and privacy of patient data by creating immutable records that can be accessed by authorized healthcare providers. This facilitates secure data sharing across different healthcare systems, enhancing the continuity of care.

- **Impact:**
  - **Improved Access:** Patients in remote areas can receive timely medical consultations without the need for travel, reducing healthcare disparities.
  - **Enhanced Monitoring:** Remote monitoring devices, integrated with AI and blockchain, provide continuous health data to clinicians, enabling proactive management of chronic conditions and timely interventions.

AI-Driven Clinical Trials: AI is transforming the design and management of clinical trials, accelerating the development of new treatments.

- **Implementation:**
  - **Patient Recruitment:** AI algorithms analyse medical records and genetic data to identify suitable candidates for clinical trials, improving the speed and accuracy of patient recruitment.
  - **Trial Management:** AI systems manage clinical trial logistics, monitor patient adherence, and analyse trial data in real-time, identifying potential issues early and ensuring the integrity of the trial.
- **Impact:**
  - **Faster Drug Development:** Streamlined trial processes reduce the time required to bring new

treatments to market, benefiting patients and healthcare providers.
- **Improved Outcomes:** AI-driven insights enhance the design of clinical trials, increasing the likelihood of successful outcomes and reducing costs.

5.6.2 Finance Deep Dive

Blockchain in Banking: Major banks are adopting blockchain technology to streamline operations and reduce costs, transforming the financial industry.

- **Implementation:**
  - **Transaction Processing:** Blockchain enables secure, real-time processing of transactions, reducing the need for intermediaries and lowering transaction costs. For example, JPMorgan Chase uses its Quorum blockchain platform to facilitate interbank transfers.
  - **Smart Contracts:** Banks use smart contracts to automate complex financial agreements, such as loan processing and trade finance, ensuring transparency and reducing the risk of fraud.
- **Impact:**
  - **Operational Efficiency:** Blockchain streamlines banking operations, reducing processing times and operational costs while enhancing security and transparency.
  - **Customer Trust:** The immutable nature of blockchain records builds customer trust, as all transactions are verifiable and tamper-proof.

Quantum Computing in Risk Management: Financial institutions are leveraging quantum computing to enhance risk management and portfolio optimization.

- **Implementation:**

- **Risk Analysis:** Quantum algorithms analyse vast datasets to identify risk factors and correlations that classical computers cannot detect. This enables more accurate risk assessments and mitigation strategies.
- **Portfolio Optimization:** Quantum computing optimizes investment portfolios by solving complex mathematical problems related to asset allocation and diversification, maximizing returns while minimizing risk.

- **Impact:**
  - **Enhanced Risk Management:** Quantum-driven insights improve the ability of financial institutions to manage and mitigate risks, ensuring greater stability and resilience.
  - **Optimized Investments:** More efficient portfolio management leads to better investment outcomes, benefiting both financial institutions and their clients.

5.6.3 Urban Planning Deep Dive

Smart Transportation Systems: AI and blockchain are key to developing smart transportation systems that reduce traffic congestion and improve urban mobility.

- **Implementation:**
  - **AI for Traffic Management:** AI analyses real-time traffic data to optimize traffic light timings, reroute traffic, and predict congestion patterns, improving traffic flow and reducing delays.
  - **Blockchain for Mobility Services:** Blockchain ensures the transparency and security of ride-sharing and public transportation services, facilitating seamless payments and data sharing among different service providers.
- **Impact:**

- **Reduced Congestion:** Smart traffic management systems lead to smoother traffic flow, reducing congestion and emissions.
- **Improved Mobility:** Enhanced coordination of transportation services increases the efficiency and convenience of urban mobility, benefiting residents and visitors.

Sustainable Urban Development: Quantum computing aids in modeling and implementing sustainable urban development practices, reducing environmental impact.

- **Implementation:**
  - **Urban Simulations:** Quantum computing simulates the environmental and social impacts of various urban development scenarios, considering factors such as energy use, water management, and waste reduction.
  - **Sustainable Design:** Quantum-enhanced simulations inform the design of green buildings and infrastructure, optimizing resource use and minimizing environmental footprints.
- **Impact:**
  - **Sustainability:** Quantum-driven insights help cities implement sustainable development practices, reducing their carbon footprint and promoting environmental stewardship.
  - **Resilience:** More accurate modeling of urban systems enhances resilience to environmental changes and disasters, ensuring long-term sustainability.

5.6.4 Public Policy Deep Dive

AI for Social Welfare Programs: AI is optimizing social welfare programs, ensuring efficient resource allocation and better outcomes for beneficiaries.

- **Implementation:**
  - **Resource Allocation:** AI algorithms analyse demographic and economic data to identify areas of need and allocate resources more effectively. For instance, AI can prioritize funding for housing assistance in regions with high homelessness rates.
  - **Program Management:** AI monitors and evaluates the effectiveness of social welfare programs, providing real-time feedback and recommendations for improvement.
- **Impact:**
  - **Efficiency:** Optimized resource allocation ensures that social welfare programs reach those most in need, maximizing their impact.
  - **Better Outcomes:** Data-driven insights improve program design and implementation, leading to better outcomes for beneficiaries.

Blockchain for Citizen Services: Blockchain enhances citizen services by providing secure and transparent access to government services.

- **Implementation:**
  - **Digital Identity:** Blockchain-based digital identity systems ensure secure and convenient access to government services, such as voting, tax filing, and social benefits. Estonia's e-Residency program is a prime example of blockchain-enhanced citizen services.
  - **Service Delivery:** Blockchain records all interactions between citizens and government agencies, ensuring transparency and accountability in service delivery.
- **Impact:**

- **Security:** Blockchain protects citizen data from tampering and unauthorized access, enhancing trust in government services.
- **Transparency:** Transparent records build public trust in government operations, as all transactions and decisions are verifiable and accountable.

Conclusion: The integration of quantum computing, AI, and blockchain technology is transforming healthcare, finance, urban planning, and public policy. Through detailed case studies and practical examples, this section illustrates how these technologies are driving innovation, efficiency, and sustainability in various industries. As these technologies continue to evolve, their impact will only grow, leading to more connected, efficient, and resilient societies.

This comprehensive exploration of industry-specific deep dives highlights the transformative potential of integrating quantum computing, AI, and blockchain technologies. By examining specific examples and case studies, readers can appreciate the profound impact these technologies are having on healthcare, finance, urban planning, and public policy.

Chapter 6: Ethical and Societal Implications

6.1 Ethical Considerations of Advanced Technologies

As quantum computing, artificial intelligence (AI), and blockchain technology converge to reshape industries and everyday life, it is crucial to address the ethical considerations associated with these advancements. This chapter explores the foundational principles of tech ethics, specific ethical concerns related to AI, quantum computing, and blockchain, and strategies to ensure these technologies are developed and deployed responsibly.

6.1.1 Foundations of Tech Ethics

Defining Tech Ethics: Tech ethics refers to the principles and standards that guide the responsible development and use of technology. It encompasses considerations such as privacy, security, fairness, transparency, and accountability, aiming to ensure that technological advancements benefit society as a whole without causing harm.

- **Importance:** Tech ethics provides a framework for identifying and addressing potential risks and negative impacts of new technologies. By adhering to ethical principles, developers and policymakers can build trust and foster public acceptance of technological innovations.
- **Principles:** Key principles of tech ethics include:
    - **Beneficence:** Ensuring that technology benefits individuals and society.
    - **Non-maleficence:** Preventing harm caused by technology.
    - **Autonomy:** Respecting the autonomy of individuals in their interactions with technology.
    - **Justice:** Ensuring fair distribution of the benefits and burdens of technology.

Historical Perspective: Ethical dilemmas posed by past technological advancements provide valuable lessons for addressing current and future challenges.

- **Case Study: Nuclear Technology**
    - **Ethical Dilemma:** The development of nuclear technology posed significant ethical challenges related to its dual use for energy production and weapons of mass destruction.
    - **Lessons Learned:** The need for robust international agreements, regulatory frameworks, and ethical guidelines to prevent misuse and ensure safe and beneficial use of technology.

- **Case Study: The Internet and Data Privacy**
    - **Ethical Dilemma:** The rise of the internet led to concerns about data privacy and security, with personal information becoming increasingly vulnerable to misuse.
    - **Lessons Learned:** The importance of developing strong data protection laws, transparency in data collection practices, and empowering individuals with control over their personal data.

6.1.2 AI Ethics

Bias and Fairness: AI algorithms can inadvertently perpetuate biases present in training data, leading to unfair or discriminatory outcomes.

- **Sources of Bias:** Bias in AI can arise from various sources, including biased training data, flawed algorithmic design, and human biases introduced during development.
- **Impact on Fairness:** Biased AI systems can disproportionately affect marginalized communities, leading to unequal access to opportunities and services.
- **Mitigation Strategies:**
    - **Diverse Datasets:** Ensuring that training data is representative of diverse populations.

- **Fairness Audits:** Conducting regular audits to identify and address biases in AI systems.
- **Algorithmic Transparency:** Making the decision-making processes of AI systems transparent to enable scrutiny and accountability.

Transparency and Explainability: Ensuring that AI decision-making processes are transparent and explainable is crucial for accountability and trust.

- **Importance:** Transparent and explainable AI allows stakeholders to understand how decisions are made, enabling them to identify and rectify errors or biases.
- **Strategies:**
  - **Explainable AI (XAI):** Developing AI models that provide clear explanations of their outputs and decision-making processes.
  - **User Education:** Educating users about how AI systems work and the factors influencing their decisions.

Autonomy and Control: The rise of autonomous systems raises ethical considerations around the need for human oversight and control.

- **Ethical Concerns:** Autonomous systems, such as self-driving cars and AI-powered weapons, can operate without direct human intervention, raising concerns about accountability and the potential for unintended consequences.
- **Human Oversight:** Ensuring that autonomous systems are subject to human oversight and intervention is crucial for preventing harm and maintaining control over technology.

6.1.3 Quantum Computing Ethics

Data Security and Quantum Cryptography: Quantum computing poses significant ethical implications for data security, particularly the potential to break current cryptographic systems.

- **Cryptographic Vulnerability:** Quantum computers could theoretically break widely used cryptographic algorithms, such as RSA and ECC, compromising the security of digital communications and transactions.
- **Ethical Implications:** The potential for quantum computing to undermine data security necessitates the development of quantum-resistant cryptographic methods to protect sensitive information.
- **Strategies:**
    - **Quantum-Resistant Algorithms:** Investing in research and development of new cryptographic algorithms that can withstand quantum attacks.
    - **Transition Planning:** Developing plans for transitioning to quantum-resistant encryption methods across industries and government sectors.

Access and Equity: Ensuring equitable access to quantum computing resources is essential to prevent the concentration of power and knowledge in the hands of a few.

- **Equitable Access:** Quantum computing resources should be accessible to a diverse range of stakeholders, including academic institutions, startups, and underrepresented communities.
- **Preventing Concentration of Power:** Measures should be taken to prevent large corporations and wealthy nations from monopolizing quantum computing capabilities.
- **Strategies:**
    - **Public Funding:** Supporting public funding initiatives for quantum research and development to democratize access.
    - **Collaboration:** Promoting international collaboration and knowledge-sharing to ensure broad access to quantum technologies.

6.1.4 Blockchain Ethics

Decentralization and Trust: Blockchain technology's decentralization poses unique ethical considerations, including the challenges of trust and governance in a decentralized world.

- **Trust Issues:** While blockchain aims to create trustless systems, the trustworthiness of the entities involved, such as developers and validators, remains crucial.
- **Governance:** Decentralized governance models, such as Decentralized Autonomous Organizations (DAOs), raise questions about accountability, decision-making processes, and the potential for abuse.
- **Strategies:**
    - **Transparent Governance:** Ensuring transparency in governance models to build trust among participants.
    - **Accountability Mechanisms:** Developing mechanisms to hold participants accountable for their actions within decentralized systems.

Privacy and Anonymity: Balancing privacy with the need for accountability in blockchain systems is a significant ethical challenge.

- **Privacy Concerns:** Blockchain's transparency can conflict with the need for privacy, exposing sensitive information.
- **Anonymity Issues:** While anonymity can protect user privacy, it can also facilitate illicit activities.
- **Strategies:**
    - **Zero-Knowledge Proofs:** Implementing zero-knowledge proofs and other privacy-preserving techniques to enhance privacy while maintaining accountability.
    - **Regulatory Compliance:** Ensuring that blockchain systems comply with data protection regulations and balance privacy with transparency.

Conclusion: The ethical considerations associated with advanced technologies such as AI, quantum computing, and blockchain are complex and multifaceted. By understanding the foundations of tech ethics and addressing specific ethical challenges, stakeholders can ensure that these technologies are developed and deployed responsibly. Balancing innovation with ethical principles is crucial for maximizing the benefits of these technologies while minimizing potential harms. As we navigate the future of technology, a commitment to ethical considerations will be essential for fostering trust, equity, and societal well-being.

This comprehensive exploration of ethical considerations highlights the importance of integrating ethical principles into the development and use of advanced technologies. By examining specific examples and strategies, readers can appreciate the critical role ethics plays in guiding the responsible evolution of AI, quantum computing, and blockchain technologies.

6.2 Impacts on Jobs

The convergence of quantum computing, artificial intelligence (AI), and blockchain technology is set to transform the job market significantly. This transformation will have profound implications for employment, skills requirements, and workforce dynamics. This section delves into the impacts on jobs, including automation and job displacement, the need for reskilling and workforce transformation, and the emergence of new job roles.

6.2.1 Automation and Job Displacement

Impact of AI and Automation: AI and automation are reshaping various industries by increasing efficiency and productivity, but they also pose the risk of job displacement.

- **Manufacturing:** Automation has already revolutionized manufacturing with the advent of robotics and AI-driven processes. Tasks that were once manual and labor-intensive are now automated, leading to significant reductions in the need for human labor.
    - **Example:** Car manufacturing plants increasingly use robotic arms for tasks such as welding, painting, and assembly, reducing the number of factory workers required on the production line.
- **Services:** The service industry is also experiencing the effects of automation. AI-powered chatbots, virtual assistants, and automated customer service systems are replacing human jobs in areas like customer support and information management.
    - **Example:** AI chatbots like those used by large telecom companies handle customer inquiries and troubleshoot issues, reducing the need for large customer service teams.

Case Studies:

- **Retail:** Retail giants like Amazon are deploying AI and automation in their logistics and supply chain operations. Automated warehouses use robots to pick, pack, and ship products, significantly reducing the need for human warehouse workers.
    - **Impact:** While automation increases efficiency and reduces costs, it also leads to job losses in traditional retail and warehousing roles.
- **Finance:** In the finance sector, AI algorithms perform tasks such as data analysis, risk assessment, and even trading. Automated trading systems can execute trades faster and more accurately than human traders.

- **Impact:** This automation leads to job displacement for traditional financial analysts and traders but also opens up new opportunities for tech-savvy professionals.

6.2.2 *Reskilling and Workforce Transformation*

Need for Reskilling: As automation and AI displace traditional jobs, the importance of reskilling and upskilling workers to prepare them for new roles in an AI-driven economy cannot be overstated.

- **Workforce Adaptation:** Workers must adapt to the changing job landscape by acquiring new skills relevant to emerging technologies. This requires a concerted effort from individuals, companies, and governments to invest in education and training programs.

Training Programs:

- **Government Initiatives:** Various governments are launching initiatives to support workforce reskilling. For example, Singapore's SkillsFuture program provides citizens with credits to pursue courses in emerging technologies, helping them stay competitive in the job market.

    - **Example:** SkillsFuture offers courses in AI, data analytics, and blockchain, equipping workers with the skills needed for the future economy.

- **Corporate Efforts:** Companies are also playing a crucial role in reskilling their employees. Tech giants like IBM and Google have launched training programs to help workers transition into new roles.

    - **Example:** IBM's P-TECH program partners with schools to provide students with the education and skills needed for technology jobs, ensuring a pipeline of qualified workers for the future.

6.2.3 Emerging Job Roles

New Opportunities: The rise of advanced technologies is not only displacing jobs but also creating new opportunities. Several new job roles are emerging, requiring skills that align with these technologies.

- **AI Ethics Officers:** As the ethical implications of AI become more prominent, there is a growing demand for professionals who can ensure that AI systems are developed and deployed responsibly.
    - **Role:** AI ethics officers are responsible for creating guidelines, conducting audits, and ensuring compliance with ethical standards in AI projects.

- **Quantum Software Developers:** The development of quantum computing technology is creating a need for specialized software developers who can create algorithms and applications for quantum computers.
    - **Role:** Quantum software developers design and implement software that leverages the unique capabilities of quantum computing to solve complex problems.

- **Blockchain Architects:** As blockchain technology becomes more widespread, there is an increasing demand for experts who can design and implement blockchain solutions.
    - **Role:** Blockchain architects are responsible for developing blockchain frameworks, creating smart contracts, and ensuring the security and scalability of blockchain systems.

Skills in Demand: To thrive in the evolving job market, workers need to acquire specific skills that will be in high demand.

- **Technical Skills:** Proficiency in AI, machine learning, quantum computing, and blockchain technologies will be crucial. This includes knowledge of programming languages such as Python, R, and Solidity.
- **Analytical Skills:** The ability to analyse large datasets and derive meaningful insights is essential, particularly in roles involving AI and data science.
- **Soft Skills:** In addition to technical skills, soft skills such as critical thinking, problem-solving, and adaptability will be highly valued. These skills enable workers to navigate the complexities of the modern job market and adapt to rapid technological changes.

Conclusion: The convergence of quantum computing, AI, and blockchain is reshaping the job market, leading to both challenges and opportunities. While automation and AI are displacing traditional jobs, the need for reskilling and workforce transformation is creating new pathways for employment. Emerging job roles in AI ethics, quantum software development, and blockchain architecture highlight the evolving nature of work. By understanding these changes and acquiring relevant skills, individuals can thrive in the future job market, ensuring that technological advancements benefit society as a whole.

This comprehensive exploration of the impacts on jobs highlights the transformative potential of integrating quantum computing, AI, and blockchain technologies. By examining specific examples and strategies, readers can appreciate the profound impact these technologies are having on employment and workforce dynamics.

6.3 Impacts on Privacy

As quantum computing, artificial intelligence (AI), and blockchain technology converge, they bring about significant changes in how data is collected, managed, and protected. These advancements raise important privacy concerns that need to be addressed to ensure that technological progress does not come at the cost of individual privacy. This section explores the impacts on privacy, focusing on data collection and surveillance, personal data ownership, and regulatory frameworks.

6.3.1 Data Collection and Surveillance

Data Privacy Concerns: The massive data collection enabled by AI and blockchain technologies poses significant privacy risks. AI systems rely on vast amounts of data to function effectively, while blockchain technology's transparency can inadvertently expose sensitive information.

- **Examination of Privacy Concerns:**
    - **AI Systems:** AI systems collect and analyse data from various sources, including social media, online transactions, and IoT devices. This extensive data collection can lead to privacy breaches if not managed properly. For instance, AI algorithms can infer sensitive information about individuals from seemingly innocuous data, such as purchasing patterns or social media activity.
    - **Blockchain Transparency:** Blockchain's immutable and transparent nature means that once data is recorded, it cannot be altered or deleted. While this ensures data integrity, it also means that personal information recorded on a blockchain can be permanently visible, raising privacy concerns.

Surveillance: Increased surveillance capabilities enabled by AI and other technologies have significant ethical implications, particularly concerning the balance between security and privacy.

- **Discussion on Ethical Implications:**

- **Enhanced Surveillance:** AI-powered surveillance systems, such as facial recognition and predictive policing, can monitor and analyse individual behavior in real-time. While these technologies can enhance security and prevent crime, they also pose significant privacy risks and can lead to state overreach and abuse.
- **Balancing Security and Privacy:** Finding the right balance between security and privacy is crucial. Excessive surveillance can lead to a loss of privacy and civil liberties, while inadequate surveillance may compromise public safety. Ethical considerations must guide the deployment of surveillance technologies to protect individual privacy rights.

6.3.2 Personal Data Ownership

Data Sovereignty: Data sovereignty refers to the concept that individuals should have the right to own and control their personal data. This principle is becoming increasingly important as data becomes a valuable asset in the digital age.

- **The Concept of Data Sovereignty:**
  - **Individual Control:** Data sovereignty empowers individuals to control who can access their data, how it is used, and for what purposes. This control extends to the ability to withdraw consent and demand the deletion of personal data.
  - **Right to Privacy:** Upholding data sovereignty is essential for protecting the right to privacy in an increasingly digital world. Individuals must be able to trust that their personal information is handled responsibly and ethically.

Blockchain for Privacy: Blockchain technology can play a pivotal role in enabling individuals to control and protect their personal data, supporting the principle of data sovereignty.

- **How Blockchain Enables Privacy:**
    - **Decentralized Identity Management:** Blockchain can facilitate decentralized identity management systems, where individuals control their digital identities and personal information. For example, solutions like Sovrin and uPort allow users to manage their identities and share specific data points only when necessary.
    - **Privacy-Preserving Technologies:** Blockchain can integrate privacy-preserving technologies such as zero-knowledge proofs (ZKPs) and homomorphic encryption. ZKPs allow verification of data without revealing the data itself, enabling secure and private transactions.
    - **Data Ownership and Consent:** Blockchain can record and enforce data ownership and consent agreements, ensuring that individuals' preferences regarding data use are respected and enforced across different platforms and services.

6.3.3 Regulatory Frameworks

Privacy Regulations: Global privacy regulations such as the General Data Protection Regulation (GDPR) and the California Consumer Privacy Act (CCPA) play a critical role in shaping the development and deployment of new technologies.

- **Overview of Global Privacy Regulations:**
    - **GDPR:** The GDPR sets stringent requirements for data protection and privacy in the European Union. It gives individuals greater control over their personal data, including rights to access, correct, and delete data, as well as to object to data processing.
    - **CCPA:** The CCPA provides similar protections for residents of California, including the right to know what personal data is being collected, the right to

request deletion of personal data, and the right to opt-out of the sale of personal data.

Future Directions: Privacy regulations will need to evolve in response to advancements in AI, quantum computing, and blockchain to address new challenges and ensure robust data protection.

- **Predictions on the Evolution of Privacy Regulations:**
    - **Adapting to AI and Quantum Computing:** As AI and quantum computing technologies evolve, privacy regulations will need to address the unique challenges they pose. This includes ensuring AI transparency, mitigating bias, and developing quantum-resistant encryption standards to protect data against potential quantum threats.
    - **Global Harmonization:** There may be a move towards greater harmonization of privacy regulations globally to provide consistent protection for individuals' data regardless of jurisdiction. This could involve international agreements and standards-setting bodies working together to create cohesive frameworks.
    - **Emphasis on Data Sovereignty:** Future regulations are likely to place greater emphasis on data sovereignty, ensuring that individuals have control over their personal data and that their privacy rights are upheld in the digital age.

Conclusion: The convergence of quantum computing, AI, and blockchain technology presents both opportunities and challenges for data privacy. While these technologies can enhance data management and protection, they also raise significant privacy concerns that need to be addressed. By understanding the impacts on data collection and surveillance, personal data ownership, and regulatory frameworks, stakeholders can develop strategies to protect individual privacy in the digital age. Balancing technological innovation with robust privacy protections is crucial for building trust and ensuring that advancements benefit society as a whole.

This comprehensive exploration of the impacts on privacy highlights the importance of addressing privacy concerns in the development and use of advanced technologies. By examining specific examples and strategies, readers can appreciate the critical role of privacy in guiding the responsible evolution of AI, quantum computing, and blockchain technologies.

6.4 Impacts on Societal Structures

The convergence of quantum computing, artificial intelligence (AI), and blockchain technology is not only reshaping industries but also profoundly impacting societal structures. This section explores the effects on social inequality, governance and democracy, and cultural and ethical shifts, offering a comprehensive view of how these technologies can transform our world.

6.4.1 Social Inequality

Technology and Inequality: Advanced technologies have the potential to exacerbate social inequalities if not managed properly. Access to and the benefits derived from these technologies can create significant disparities between different social groups.

- **Examination of Inequality:**
    - **Economic Disparities:** Wealthy individuals and nations are more likely to access and benefit from cutting-edge technologies, leading to increased economic disparities. For instance, companies with substantial resources can invest in AI and quantum computing to gain a competitive edge, leaving smaller businesses and developing countries at a disadvantage.

- **Education and Skills Gap:** Access to quality education and training in advanced technologies is unevenly distributed. Individuals with access to better educational resources are more likely to acquire the skills needed for high-paying tech jobs, further widening the income gap.

Digital Divide: The digital divide refers to the gap between those who have access to modern information and communication technology and those who do not. This divide has significant implications for equitable access to technology.

- **Implications for Equitable Access:**
    - **Infrastructure Disparities:** Many rural and underserved areas lack the necessary infrastructure to support high-speed internet and other technological advancements, limiting residents' access to digital services and opportunities.
    - **Inclusion Initiatives:** Bridging the digital divide requires targeted initiatives to expand infrastructure, provide affordable access to technology, and deliver digital literacy programs. Governments and private sector partnerships can play a crucial role in addressing these disparities.

6.4.2 Governance and Democracy

Decentralized Governance: Blockchain technology has the potential to enable new forms of decentralized governance and direct democracy, shifting power dynamics and increasing citizen participation.

- **Potential for Decentralized Governance:**
    - **Transparency and Trust:** Blockchain's transparent and immutable ledger can enhance trust in governance by ensuring that all transactions and decisions are publicly verifiable and tamper-proof.

This transparency can reduce corruption and increase accountability.
- **Direct Democracy:** Blockchain can facilitate direct democracy by enabling secure and transparent voting systems. Citizens can vote on policies and decisions directly, bypassing traditional bureaucratic processes.

Policy Making with AI: AI can support data-driven policymaking by analysing large datasets to provide insights and predictions, while ensuring transparency and accountability.

- **Supporting Data-Driven Policy Making:**
  - **Predictive Analytics:** AI can analyse social, economic, and environmental data to predict the outcomes of various policy decisions. This allows policymakers to make more informed choices that are based on empirical evidence.
  - **Transparency and Accountability:** AI systems must be designed with transparency in mind. Explainable AI (XAI) ensures that decision-making processes are understandable, enabling policymakers and the public to scrutinize and trust AI-generated recommendations.

The Impact of the MOTMSDD Approach in Public Decision Making: The Metaverse of the Minds Social Direct Democracy (MOTMSDD) approach leverages advanced technologies to enhance public engagement and decision-making processes.

- **Implementation and Impact:**
  - **Digital Twin Representation:** In the MOTMSDD metaverse, each individual has a digital twin representing their preferences and needs. These digital twins participate in decision-making processes in real-time, ensuring that public policies reflect the true needs of the population.

- **Enhanced Public Participation:** MOTMSDD integrates brain-computer interfaces (BCIs) and AI to facilitate seamless public engagement. Citizens can express their views and vote on policies directly, making the decision-making process more democratic and inclusive.
- **Conflict Resolution:** AI and quantum computing within the MOTMSDD framework can resolve conflicts between stakeholders by simulating various scenarios and finding optimal solutions that balance competing interests.

6.4.3 Cultural and Ethical Shifts

Changing Norms: Advanced technologies are changing societal norms and ethical standards, influencing how we interact, work, and live.

- **Shifts in Societal Norms:**
  - **Work and Automation:** The automation of tasks and the rise of remote work are reshaping traditional work norms. As AI and robots take over routine tasks, there is a growing emphasis on creativity, problem-solving, and emotional intelligence in the workplace.
  - **Privacy and Surveillance:** The proliferation of surveillance technologies and data collection practices is altering norms around privacy. Societies are grappling with the balance between security and individual privacy rights.

Public Perception: Public perception of advanced technologies is crucial for their acceptance and integration into everyday life. Fostering trust and understanding is essential for overcoming skepticism and resistance.

- **Discussion on Public Perception:**

- **Building Trust:** Transparency, accountability, and ethical considerations are key to building public trust in advanced technologies. Clear communication about the benefits and risks of these technologies can help mitigate fears and misconceptions.
- **Education and Engagement:** Public education and engagement initiatives can demystify advanced technologies and highlight their potential benefits. Involving communities in discussions about technology deployment can also ensure that their concerns are addressed and their voices heard.

Conclusion: The convergence of quantum computing, AI, and blockchain technology is profoundly impacting societal structures, including social inequality, governance, and cultural norms. By understanding and addressing the challenges and opportunities these technologies present, we can work towards a more equitable and inclusive future. Ensuring that these advancements benefit all members of society, fostering transparency and trust in governance, and adapting to changing cultural and ethical standards are crucial steps in navigating the societal impacts of these transformative technologies.

This comprehensive exploration of the impacts on societal structures highlights the transformative potential of integrating quantum computing, AI, and blockchain technologies. By examining specific examples and strategies, readers can appreciate the profound impact these technologies are having on social inequality, governance, and cultural norms.

6.5 Case Studies and Real-World Examples

Understanding the practical applications and ethical considerations of advanced technologies is essential for grasping their full impact on society. This section presents detailed case studies and real-world examples to illustrate how AI, quantum computing, and blockchain are being implemented across various sectors and the ethical challenges they pose.

6.5.1 AI Ethics in Practice

Healthcare: AI-driven healthcare solutions offer significant benefits but also present ethical challenges, particularly concerning privacy, bias, and the transparency of AI systems.

- **Case Study: AI in Diagnostic Imaging**
    - **Implementation:** Hospitals and clinics are increasingly using AI to assist in diagnostic imaging, such as interpreting X-rays, CT scans, and MRIs. AI algorithms analyse images to detect anomalies that may indicate conditions like cancer, fractures, or infections.
    - **Ethical Challenges:** The primary ethical concerns include ensuring patient privacy, addressing potential biases in AI algorithms, and maintaining transparency in AI decision-making processes.
    - **Solutions:**
        - **Privacy Protections:** Implementing robust data protection measures to secure patient information.
        - **Bias Mitigation:** Using diverse and representative training datasets to minimize bias in AI models.
        - **Transparency:** Developing explainable AI systems that provide clear rationales for their diagnoses, allowing healthcare professionals to understand and trust AI recommendations.

Criminal Justice: The use of AI in criminal justice systems raises critical ethical issues related to bias, fairness, and accountability.

- **Case Study: Predictive Policing**
    - **Implementation:** Some police departments use predictive policing algorithms to forecast where crimes are likely to occur, allowing for more targeted deployment of officers.
    - **Ethical Challenges:** Concerns include the potential for AI to reinforce existing biases, lack of transparency in algorithmic decision-making, and accountability for AI-driven actions.
    - **Solutions:**
        - **Bias Audits:** Conducting regular audits to identify and address biases in predictive policing algorithms.
        - **Transparency and Oversight:** Ensuring that the criteria and data used by predictive algorithms are transparent and subject to oversight by independent bodies.
        - **Community Engagement:** Involving community stakeholders in the development and deployment of predictive policing technologies to build trust and ensure that the systems are used fairly.

6.5.2 Quantum Computing and Society

National Security: Quantum computing holds significant implications for national security, particularly concerning cryptography and data protection.

- **Case Study: Quantum Cryptography in National Defense**
    - **Implementation:** Governments are investing in quantum cryptography to secure military communications and protect sensitive information from potential quantum attacks.
    - **Ethical Considerations:** The development and deployment of quantum cryptography raise ethical

questions about the balance between national security and individual privacy, as well as the potential for an arms race in quantum technologies.

- **Solutions:**
    - **International Collaboration:** Promoting international agreements and collaborations to prevent a quantum arms race and ensure the peaceful use of quantum technologies.
    - **Ethical Guidelines:** Developing ethical guidelines for the use of quantum cryptography in national defense, emphasizing respect for privacy and human rights.

Scientific Research: Quantum computing offers unparalleled potential for scientific research but also raises ethical concerns, particularly regarding dual-use technologies.

- **Case Study: Quantum Simulations in Drug Discovery**
    - **Implementation:** Pharmaceutical companies use quantum computers to simulate molecular interactions at unprecedented scales, accelerating the discovery of new drugs.
    - **Ethical Considerations:** Ethical concerns include the potential dual-use of quantum technologies for harmful purposes, such as creating new chemical or biological weapons.
    - **Solutions:**
        - **Regulatory Oversight:** Implementing stringent regulatory frameworks to monitor and control the use of quantum computing in sensitive research areas.
        - **Ethical Review Boards:** Establishing ethical review boards to assess the potential risks and benefits of quantum research projects.

6.5.3 Blockchain for Social Good

Supply Chain Transparency: Blockchain technology can enhance transparency and ethics in supply chains, ensuring that products are sourced responsibly and ethically.

- **Case Study: Blockchain in the Food Industry**
    - **Implementation:** Companies like IBM and Walmart use blockchain to track the provenance of food products from farm to table, ensuring that they are ethically sourced and safe for consumption.
    - **Ethical Implications:** The use of blockchain for supply chain transparency raises issues related to data privacy and the potential exclusion of smaller suppliers who may lack the resources to adopt blockchain technology.
    - **Solutions:**
        - **Inclusive Design:** Ensuring that blockchain solutions are designed to be accessible to all participants in the supply chain, including small and medium-sized enterprises (SMEs).
        - **Data Privacy Protections:** Implementing privacy-preserving techniques to protect sensitive business information while maintaining transparency.

Voting Systems: Blockchain can provide secure and transparent voting systems, addressing issues of fraud and disenfranchisement but also raising ethical concerns.

- **Case Study: Blockchain-Based Voting in Estonia**
    - **Implementation:** Estonia has implemented a blockchain-based e-voting system that allows citizens to vote securely and transparently in national elections.

- **Ethical Implications:** While blockchain can enhance the security and transparency of voting systems, it also raises concerns about voter privacy, the digital divide, and the potential for coercion.
- **Solutions:**
    - **Privacy Enhancements:** Using privacy-preserving technologies, such as zero-knowledge proofs, to protect voter privacy while ensuring the integrity of the vote.
    - **Digital Inclusion:** Implementing programs to ensure that all citizens have access to the technology needed to participate in blockchain-based voting systems.
    - **Coercion Prevention:** Developing measures to prevent voter coercion and ensure that the voting process remains free and fair.

Conclusion: These case studies and real-world examples illustrate the diverse applications and ethical considerations of AI, quantum computing, and blockchain technologies. By examining these examples, stakeholders can better understand the practical challenges and opportunities presented by these technologies. Addressing the ethical implications and ensuring that technological advancements are used responsibly will be crucial for maximizing their benefits and minimizing potential harms.

This comprehensive exploration of case studies and real-world examples highlights the transformative potential of integrating quantum computing, AI, and blockchain technologies. By examining specific instances and ethical challenges, readers can appreciate the practical impact of these technologies on various sectors and the importance of addressing ethical considerations in their development and deployment.

6.6 Preparing for the Future

The convergence of quantum computing, artificial intelligence (AI), and blockchain technology is poised to transform industries and everyday life in profound ways over the next decade. To ensure these advancements benefit society while mitigating potential risks, it is crucial to prepare by building ethical frameworks, enhancing education and public awareness, and fostering global collaboration. This section outlines strategies for preparing for the future, focusing on ethical guidelines, education, public engagement, and international cooperation.

6.6.1 Building Ethical Frameworks

Ethical Guidelines: Developing robust ethical guidelines is essential for the responsible use of advanced technologies. These guidelines should be comprehensive, covering all stages of technology development and deployment.

- **Development of Ethical Guidelines:**
    - **Stakeholder Involvement:** Involving diverse stakeholders, including ethicists, technologists, policymakers, and representatives from affected communities, ensures that ethical guidelines are well-rounded and consider various perspectives.
    - **Dynamic and Adaptive:** Ethical guidelines must be dynamic and adaptable to keep pace with rapid technological advancements. Regular reviews and updates are necessary to address emerging ethical challenges.

Industry Standards: Industry standards and best practices play a critical role in ensuring the ethical development and deployment of advanced technologies.

- **The Role of Industry Standards:**

- **Consistency and Compliance:** Industry standards provide a consistent framework for ethical technology development, ensuring that companies adhere to best practices. Compliance with these standards can be incentivized through certifications and accreditation.
- **Collaboration and Sharing:** Developing industry standards requires collaboration among companies, regulatory bodies, and standards organizations. Sharing best practices and lessons learned helps create a robust ethical framework that benefits the entire industry.

6.6.2 Education and Public Awareness

Ethics Education: Incorporating ethics education into technology-related curricula is vital for preparing future technologists to navigate the ethical challenges associated with advanced technologies.

- **Importance of Ethics Education:**
    - **Curriculum Integration:** Ethics education should be integrated into all levels of technology education, from primary school to higher education and professional training programs. This ensures that ethical considerations are ingrained in the minds of future technologists from an early age.
    - **Interdisciplinary Approach:** Combining technical education with courses in philosophy, sociology, and law helps students understand the broader societal implications of their work. This interdisciplinary approach fosters critical thinking and ethical reasoning.

Public Engagement: Engaging the public in discussions about the ethical implications of advanced technologies is crucial for building trust and ensuring that these technologies are developed in a way that aligns with societal values.

- **Strategies for Public Engagement:**
    - **Community Forums and Workshops:** Organizing community forums, workshops, and town hall meetings provides a platform for public discussion and input on ethical issues related to advanced technologies.
    - **Public Awareness Campaigns:** Using media, social networks, and public service announcements to raise awareness about the ethical implications of advanced technologies and the importance of responsible development.
    - **Citizen Science and Participatory Research:** Encouraging public participation in scientific research and technology development projects helps demystify advanced technologies and fosters a sense of ownership and accountability.

6.6.3 *Global Collaboration*

International Cooperation: Addressing the ethical and societal implications of advanced technologies requires international cooperation. Global challenges such as data privacy, cybersecurity, and equitable access to technology can only be effectively addressed through collaborative efforts.

- **The Need for International Cooperation:**
    - **Global Standards:** Developing global standards for ethical technology development ensures consistency and fairness across borders. International organizations, such as the United Nations and the World Economic Forum, can play a pivotal role in facilitating these efforts.
    - **Cross-Border Collaboration:** Governments, academic institutions, and private companies should collaborate on research and development projects to share knowledge, resources, and best practices. This collaboration can help bridge the gap between

developed and developing countries, ensuring that technological benefits are distributed more equitably.

Global Initiatives: Several global initiatives aim to promote ethical technology development and use, addressing the societal challenges posed by advanced technologies.

- **Overview of Global Initiatives:**
    - **The Global Partnership on Artificial Intelligence (GPAI):** GPAI is an international initiative that brings together experts from government, industry, and academia to promote the responsible use of AI. It focuses on areas such as AI ethics, fairness, transparency, and accountability.
    - **The International Telecommunication Union (ITU):** ITU is a specialized agency of the United Nations that works to develop international standards for information and communication technologies, including those related to AI and quantum computing.
    - **The IEEE Global Initiative on Ethics of Autonomous and Intelligent Systems:** This initiative aims to ensure that autonomous and intelligent systems are designed and implemented in a way that prioritizes human well-being and ethical considerations.

Conclusion: Preparing for the future of quantum computing, AI, and blockchain technology requires a multifaceted approach that includes building ethical frameworks, enhancing education and public awareness, and fostering global collaboration. By developing robust ethical guidelines and industry standards, integrating ethics education into technology curricula, engaging the public in meaningful discussions, and promoting international cooperation, we can ensure that these technologies are developed and deployed in a manner that benefits society as a whole. This proactive approach will help navigate the ethical and societal challenges posed by advanced technologies, fostering a more equitable, inclusive, and sustainable future.

This comprehensive exploration of preparing for the future highlights the importance of integrating ethical considerations, education, public engagement, and global collaboration into the development and deployment of advanced technologies. By examining specific strategies and initiatives, readers can appreciate the critical steps needed to ensure that quantum computing, AI, and blockchain technologies are used responsibly and beneficially.

Chapter 7: Future Predictions

7.1 Visionary Predictions on the Next 10-20 Years

As we look towards the future, the convergence of quantum computing, artificial intelligence (AI), and blockchain technology promises to reshape industries and everyday life in unprecedented ways. This section explores visionary predictions for the next 10-20 years, offering insights into the evolution of AI, quantum computing breakthroughs, the expansion of blockchain and decentralization, and the transformative potential of these converging technologies.

7.1.1 The Evolution of AI

General AI: Predictions on the development of Artificial General Intelligence (AGI) indicate that we are moving closer to creating machines that possess human-like cognitive abilities.

- **Development of AGI:** AGI will be capable of understanding, learning, and applying knowledge across a wide range of tasks, surpassing the limitations of current narrow AI systems. The journey towards AGI involves significant advancements in machine learning, neural networks, and cognitive computing.
- **Potential Capabilities:** AGI could revolutionize numerous fields, from scientific research and healthcare to robotics and education. It will be able to autonomously design experiments, develop new technologies, and provide personalized education tailored to individual learning styles.

AI Integration: AI will become deeply integrated into everyday life, enhancing both personal and professional experiences.

- **Personal Assistants:** AI-driven personal assistants will evolve into highly intelligent companions that can manage daily schedules, provide health advice, and even offer emotional support. These assistants will be seamlessly integrated into our devices and environments, making interactions with technology more natural and intuitive.
- **Autonomous Systems:** The rise of autonomous systems, including self-driving cars, drones, and robotic helpers, will transform transportation, logistics, and

home management. These systems will operate with minimal human intervention, increasing efficiency and safety.

Ethical AI: The development and implementation of ethical AI frameworks and regulations will be crucial to ensure that AI technologies benefit society and mitigate risks.

- **Rise of Ethical AI:** Ethical AI will prioritize fairness, transparency, and accountability. Regulatory bodies will establish guidelines for AI development and deployment, ensuring that AI systems are designed and used responsibly.
- **Regulations:** Governments and international organizations will collaborate to create comprehensive AI regulations that address issues such as bias, privacy, and the ethical use of AI in decision-making processes.

7.1.2 Quantum Computing Breakthroughs

Quantum Supremacy: Future milestones in achieving and surpassing quantum supremacy will mark significant advancements in computing power.

- **Achieving Quantum Supremacy:** Quantum supremacy refers to the point at which quantum computers can solve problems that are currently intractable for classical computers. Over the next decade, we will likely see quantum computers achieving this milestone more consistently across various tasks.
- **Beyond Supremacy:** As quantum computers continue to evolve, they will tackle increasingly complex problems, driving breakthroughs in fields such as cryptography, material science, and artificial intelligence.

Quantum Applications: The expansion of quantum computing applications will revolutionize various industries.

- **Cryptography:** Quantum computing will enable the development of new cryptographic techniques that are virtually unbreakable, enhancing the security of digital communications and transactions.
- **Material Science:** Quantum simulations will accelerate the discovery of new materials with unique properties, leading to advancements in technology, medicine, and energy.
- **Complex System Simulations:** Quantum computers will model complex systems more accurately than ever before, providing insights into climate change, financial markets, and biological processes.

Quantum Ecosystem: The growth of the quantum computing ecosystem will involve advancements in hardware, software development, and the cultivation of a skilled workforce.

- **Hardware Advancements:** Continued progress in quantum hardware will lead to more stable and scalable quantum computers, overcoming current limitations such as qubit coherence and error rates.
- **Software Development:** The development of quantum programming languages and frameworks will make quantum computing more accessible to researchers and developers, fostering innovation and collaboration.
- **Skilled Workforce:** Educational institutions and industry leaders will invest in training programs to develop a skilled workforce capable of advancing and applying quantum technologies.

7.1.3 Blockchain and Decentralization

Ubiquitous Blockchain: Blockchain technology will proliferate across various industries beyond finance, driving transparency, efficiency, and security.

- **Healthcare:** Blockchain will secure patient records, streamline medical billing, and ensure the authenticity of pharmaceuticals.
- **Supply Chain:** Enhanced transparency and traceability will combat fraud, improve product quality, and ensure ethical sourcing.
- **Governance:** Blockchain will enable secure, transparent voting systems and enhance public trust in governmental processes.

Decentralized Internet (Web 3.0): The development of a decentralized internet will enhance privacy, security, and user control.

- **Web 3.0:** Web 3.0 will leverage blockchain to create a decentralized web where users have greater control over their data and digital identities. This will reduce reliance on centralized platforms and mitigate issues related to data breaches and privacy violations.

Tokenized Economy: The emergence of a tokenized economy will see digital assets and cryptocurrencies becoming mainstream and widely accepted.

- **Digital Assets:** Digital assets, including cryptocurrencies and non-fungible tokens (NFTs), will become integral to the global economy, enabling new forms of ownership, investment, and value exchange.

- **Mainstream Adoption:** Businesses and governments will increasingly accept and utilize digital currencies, integrating them into financial systems and everyday transactions.

7.1.4 Convergence of Technologies

Integrated Solutions: The convergence of AI, quantum computing, and blockchain will create integrated solutions for complex global challenges.

- **Global Challenges:** Integrated technologies will address global challenges such as climate change, healthcare, and food security by providing innovative solutions that leverage the strengths of each technology.

Smart Infrastructure: The development of smart infrastructure powered by these converging technologies will lead to more efficient and sustainable cities.

- **Smart Cities:** Smart cities will utilize AI for traffic management, quantum computing for optimizing energy use, and blockchain for secure and transparent governance. This will enhance urban living by reducing congestion, lowering emissions, and improving resource management.

Personalized Services: Highly personalized services in healthcare, education, and entertainment will emerge, driven by the synergy of these technologies.

- **Healthcare:** Personalized medicine will leverage AI to analyse genetic information, quantum computing to

simulate treatment outcomes, and blockchain to secure patient data.
- **Education:** AI-driven personalized learning platforms will adapt to individual student needs, while blockchain ensures the security and verification of academic records.
- **Entertainment:** AI and quantum computing will create immersive and interactive entertainment experiences, while blockchain manages intellectual property rights and fair compensation for creators.

The MOTMSDD Approach: The Metaverse of the Minds Social Direct Democracy (MOTMSDD) approach will increase and leverage public engagement through smart participatory AI-enhanced direct democracy.

- **Public Engagement:** MOTMSDD will utilize AI and blockchain to create a digital metaverse where citizens can participate directly in decision-making processes. This approach will ensure that public policies reflect the true needs and preferences of the population.
- **Smart Participatory Democracy:** By integrating advanced technologies, MOTMSDD will enhance transparency, accountability, and efficiency in governance, fostering a more democratic and responsive society.

Conclusion: The next 10-20 years will witness remarkable advancements in AI, quantum computing, and blockchain technology. These technologies will become deeply integrated into various aspects of life, driving significant changes in industries, governance, and societal structures. By preparing for these changes and addressing ethical considerations, we can harness the full potential of these technologies to create a better, more equitable future.

This comprehensive exploration of visionary predictions highlights the transformative potential of integrating quantum computing, AI, and blockchain technologies. By examining specific advancements and their implications, readers can appreciate the profound impact these technologies will have on shaping the future.

7.2 Insights from Industry Leaders and Futurists
7.2.1 Quotes from Thought Leaders
Key Figures:

- **AI:**

    - **Dr. Andrew Ng, Co-founder of Google Brain and Coursera:**

        - "AI is the new electricity. Just as electricity transformed almost everything 100 years ago, today I actually have a hard time thinking of an industry that I don't think AI will transform in the next several years." - Source: Harvard Business Review

    - **Fei-Fei Li, Professor at Stanford University and Co-Director of the Stanford Human-Centered AI Institute:**

        - "As much as AI is showing amazing progress, AI algorithms learn from data. And data can carry biases. It is our job to ensure fairness and transparency in these systems." - Source: WIRED

- **Quantum Computing:**

    - **Dr. John Preskill, Professor of Theoretical Physics at Caltech:**
        - *"We hope that within about 10 years we will have quantum computers that can solve useful problems that are beyond the reach of classical computers."* - Source: Caltech

    - **Scott Aaronson, Professor of Computer Science at the University of Texas at Austin:**
        - *"The true promise of quantum computing lies in its potential to solve problems in chemistry, materials science, and other fields that are currently beyond our reach."* - Source: MIT Technology Review

- **Blockchain:**

    - **Vitalik Buterin, Co-founder of Ethereum:**
        - *"The great thing about blockchain is that it democratizes access. It allows people who don't trust each other to collaborate without having to rely on intermediaries."* - Source: Fortune

    - **Dr. Gavin Wood, Co-founder of Ethereum and Polkadot:**
        - *"Blockchain technology is fundamentally about establishing trust and reliability through transparency and decentralization."* - Source: CoinDesk

Pioneering Companies:

- **AI Companies:**
  - **Sundar Pichai, CEO of Google and Alphabet:**
    - *"AI is one of the most important things humanity is working on. It is more profound than, I don't know, electricity or fire."* - Source: The Verge
  - **Arvind Krishna, CEO of IBM:**
    - *"Quantum computing will enable us to solve problems that are currently unsolvable, making significant impacts on business and society."* - Source: IBM
- **Blockchain Companies:**
  - **Brad Garlinghouse, CEO of Ripple:**
    - *"Blockchain will fundamentally change how we think about trust and value exchange."* - Source: TechCrunch
  - **Catherine Coley, CEO of Binance.US:**
    - *"The future of finance is decentralized, and blockchain is at the heart of this transformation."* - Source: [Yahoo Finance](#)

7.2.2 Futurists' Perspectives

Scenario Planning: Futurists predict various scenarios for the next 10-20 years, considering different technological, economic, and social trends.

- **Optimistic Scenario:**
  - **Technological Utopia:** Advanced technologies create a world of abundance, with AI and quantum computing solving major global challenges such as climate change, disease, and poverty. Blockchain

ensures transparency and fairness in governance and economic systems.

- **Example:** *Ray Kurzweil, renowned futurist and inventor, envisions a future where human intelligence is augmented by AI, leading to exponential growth in innovation and quality of life.* - Source: <u>The Singularity Is Near</u>

- **Cautious Scenario:**

    - **Managed Growth:** Technological advancements are significant but carefully managed to mitigate risks such as job displacement, privacy concerns, and ethical dilemmas. Policies and regulations evolve to ensure that technology benefits society as a whole.

    - **Example:** *Amy Webb, futurist and founder of the Future Today Institute, advocates for proactive planning and regulation to guide the ethical development of AI and other advanced technologies.* - Source: <u>The Big Nine</u>

Disruptive Innovations: Identification of potential disruptive innovations that could reshape industries and societal structures.

- **AI-Driven Healthcare:** AI-powered diagnostic tools and personalized medicine could revolutionize healthcare, making it more efficient and accessible. AI could also play a crucial role in managing public health crises, such as pandemics.
- **Quantum Cryptography:** The development of quantum cryptography will transform data security, making current encryption methods obsolete and creating new standards for protecting sensitive information.
- **Decentralized Autonomous Organizations (DAOs):** DAOs, enabled by blockchain technology, could revolutionize organizational governance, allowing for more democratic and transparent decision-making processes.

7.2.3 Global Impact

Economic Shifts: Predictions on how the global economy will be transformed by these technologies, including shifts in job markets and economic power.

- **Shifts in Job Markets:** As AI, quantum computing, and blockchain technologies advance, job markets will undergo significant transformations. While some jobs will be displaced by automation, new roles will emerge in technology development, data analysis, and cybersecurity.
    - **Example:** *McKinsey Global Institute predicts that up to 375 million workers may need to switch occupational categories and learn new skills by 2030 due to automation and AI.* - Source: McKinsey Global Institute

- **Economic Power:** Countries that lead in developing and adopting these technologies will gain significant economic advantages. This could shift the balance of economic power, with emerging economies potentially leapfrogging established ones by leveraging advanced technologies.

Social Implications: Insights into the social implications of widespread adoption of these technologies, such as changes in social norms, education, and daily life.

- **Changes in Social Norms:** The integration of AI and automation into daily life will alter social norms around work, privacy, and human interaction. For example, the increasing use of AI in personal assistants and home automation will change how people interact with technology and each other.

- **Education:** The education system will need to adapt to prepare students for a rapidly changing job market. Emphasis on STEM (science, technology, engineering, and mathematics) education, as well as lifelong learning and reskilling programs, will be crucial.

- **Example:** *The World Economic Forum advocates for a "Reskilling Revolution," aiming to provide better education, skills, and jobs to one billion people by 2030.* - Source: World Economic Forum

- **Daily Life:** Advanced technologies will make daily life more convenient and efficient. Smart homes, personalized healthcare, and AI-driven transportation will become commonplace, enhancing quality of life but also raising questions about privacy and data security.

Conclusion: The insights from industry leaders and futurists underscore the transformative potential of quantum computing, AI, and blockchain technologies. These perspectives highlight the opportunities and challenges that lie ahead, emphasizing the need for proactive planning, ethical guidelines, and global collaboration to ensure that these technologies benefit society as a whole. By understanding the visionary predictions and preparing for the future, we can navigate the complexities of technological advancement and create a more equitable, sustainable, and prosperous world.

This comprehensive exploration of insights from industry leaders and futurists highlights the transformative potential of integrating quantum computing, AI, and blockchain technologies. By examining specific predictions and their implications, readers can appreciate the profound impact these technologies will have on shaping the future.

7.3 Technological Advancements and Their Impacts

As we delve into the next decade, the convergence of quantum computing, artificial intelligence (AI), and blockchain technology is set to revolutionize various sectors. This section explores the anticipated advancements in healthcare, finance, urban development, and education, illustrating their profound impacts on society.

7.3.1 Healthcare Revolution

Precision Medicine: The future of precision medicine will be significantly enhanced by the integration of AI and quantum computing, leading to highly personalized treatment plans tailored to individual patients' genetic profiles and medical histories.

- **AI-Driven Diagnostics:**
    - **Implementation:** AI algorithms will analyse vast amounts of medical data, including genetic information, medical records, and lifestyle factors, to diagnose diseases with high accuracy and recommend personalized treatment plans.
    - **Impact:** This approach will lead to earlier detection of diseases, more effective treatments, and improved patient outcomes.

- **Quantum Computing in Drug Discovery:**
    - **Implementation:** Quantum computers will simulate complex molecular interactions, accelerating the discovery of new drugs and reducing the time and cost of bringing them to market.

- **Impact:** This will lead to the development of more effective medications, particularly for complex diseases such as cancer and neurodegenerative disorders.

Blockchain Health Records: Universal adoption of blockchain technology for secure, interoperable health records will transform patient care and research.

- **Implementation:**
    - **Data Security:** Blockchain's immutable ledger will ensure the security and privacy of patient records, preventing unauthorized access and data breaches.
    - **Interoperability:** Blockchain will enable seamless sharing of health records across different healthcare providers, improving coordination and continuity of care.
    - **Patient Control:** Patients will have greater control over their health data, with the ability to grant or revoke access as needed.
- **Impact:** Improved data security and interoperability will enhance patient care, facilitate research, and enable more efficient healthcare systems.

7.3.2 Finance and Economy

Decentralized Finance (DeFi): The evolution of DeFi platforms will offer a wide range of financial services without the need for traditional intermediaries, democratizing access to financial products and services.

- **Implementation:**
    - **Smart Contracts:** DeFi platforms will use smart contracts to automate financial transactions, including lending, borrowing, trading, and insurance, without the need for intermediaries.
    - **Accessibility:** These platforms will be accessible to anyone with an internet connection, providing financial services to unbanked and underbanked populations.
- **Impact:** DeFi will increase financial inclusion, reduce transaction costs, and create new economic opportunities.

Quantum Financial Models: Advanced financial modeling and risk assessment using quantum computing will lead to more robust financial systems.

- **Implementation:**
    - **Portfolio Optimization:** Quantum algorithms will optimize investment portfolios by considering a vast number of variables and scenarios, improving risk management and returns.
    - **Fraud Detection:** Quantum-enhanced algorithms will detect fraudulent activities with greater accuracy, protecting financial institutions and consumers.

- **Impact:** The use of quantum computing in finance will enhance the stability and resilience of financial systems, reducing systemic risks and improving overall efficiency.

7.3.3 Urban Development and Smart Cities

Sustainable Urban Planning: The development of sustainable, efficient, and resilient urban environments will be driven by AI and quantum simulations.

- **Implementation:**
    - **AI-Powered Planning:** AI will analyse data on traffic patterns, energy consumption, and environmental factors to optimize urban planning and infrastructure development.
    - **Quantum Simulations:** Quantum computers will simulate complex urban systems, predicting the impact of various planning decisions on sustainability and resilience.
- **Impact:** These technologies will lead to more livable, efficient, and sustainable cities, reducing environmental impact and enhancing quality of life for residents.

Blockchain Governance: The implementation of blockchain-based governance systems will enable transparent, participatory urban management.

- **Implementation:**
    - **Transparent Decision-Making:** Blockchain will provide a transparent record of government

decisions and expenditures, increasing accountability and reducing corruption.
- **Participatory Governance:** Citizens will use blockchain platforms to participate in decision-making processes, such as voting on local initiatives and budgets.

- **Impact:** Blockchain governance will enhance trust in public institutions and empower citizens to take an active role in urban management.

The MOTMSDD Approach: The future real-world implementation of the Metaverse of the Minds Social Direct Democracy (MOTMSDD) will further revolutionize public engagement and decision-making.

- **Implementation:**
    - **Digital Twins:** Each citizen will have a digital twin in the MOTMSDD metaverse, representing their preferences and needs in real-time decision-making processes.
    - **Brain-Computer Interfaces (BCIs):** BCIs will enable seamless interaction between citizens and the MOTMSDD metaverse, allowing for direct participation in policy-making.
    - **Conflict Resolution:** AI and quantum computing will be used to resolve conflicts and optimize decisions, balancing competing interests and ensuring equitable outcomes.

- **Impact:** The MOTMSDD approach will create a more inclusive and participatory form of democracy, enhancing public trust and ensuring that policies reflect the true needs of the population.

7.3.4 Education and Workforce

AI in Education: Personalized learning experiences powered by AI will adapt to individual student needs and learning styles, improving educational outcomes.

- **Implementation:**

    - **Adaptive Learning Platforms:** AI-driven platforms will provide personalized lessons and feedback, adjusting to each student's progress and preferences.

    - **Predictive Analytics:** AI will predict student performance and identify areas where additional support is needed, enabling timely interventions.

- **Impact:** AI in education will lead to more effective and engaging learning experiences, reducing dropout rates and preparing students for future technological demands.

Quantum Education: The growth of quantum computing education programs will prepare the workforce for future technological demands.

- **Implementation:**

    - **Curriculum Development:** Educational institutions will develop curricula that include quantum computing, AI, and blockchain, ensuring that students acquire the necessary skills.

- **Industry Partnerships:** Collaborations between academia and industry will provide students with practical experience and access to cutting-edge technologies.

- **Impact:** Quantum education will create a skilled workforce capable of advancing and applying quantum technologies, driving innovation and economic growth.

Conclusion: The convergence of quantum computing, AI, and blockchain technology will revolutionize various sectors, from healthcare and finance to urban development and education. By understanding the potential impacts and preparing for these advancements, we can harness the full potential of these technologies to create a more equitable, sustainable, and prosperous future.

This comprehensive exploration of technological advancements and their impacts highlights the transformative potential of integrating quantum computing, AI, and blockchain technologies. By examining specific examples and their implications, readers can appreciate the profound impact these technologies will have on shaping the future.

- .

7.4 Potential Challenges and Solutions

While the convergence of quantum computing, artificial intelligence (AI), and blockchain technology offers immense potential, it also presents significant challenges. Addressing these challenges is crucial to harnessing the benefits of these technologies responsibly. This section explores the potential ethical dilemmas, regulatory and policy issues, security risks, and societal adaptation challenges, offering strategies and solutions to navigate them.

7.4.1 Ethical Dilemmas

Bias and Discrimination: AI systems are often prone to biases that reflect historical and societal prejudices, leading to unfair and discriminatory outcomes.

- **Challenges:**
    - **Data Bias:** AI models learn from historical data, which can contain biases against certain groups based on race, gender, or socioeconomic status.
    - **Algorithmic Bias:** The algorithms themselves can introduce biases, either through design choices or unintended consequences.

- **Solutions:**
    - **Diverse Data Sources:** Ensuring that AI systems are trained on diverse and representative datasets can help mitigate data bias.
    - **Algorithm Audits:** Regularly auditing AI algorithms for bias and discrimination can help identify and correct biased behaviors.
    - **Inclusive Design:** Involving diverse teams in the design and development of AI systems can help ensure that different perspectives are considered, reducing the risk of bias.

Privacy Concerns: The advanced data collection and processing capabilities of AI and blockchain raise significant privacy concerns.

- **Challenges:**
    - **Data Surveillance:** AI systems can analyse vast amounts of personal data, leading to concerns about surveillance and loss of privacy.
    - **Blockchain Transparency:** While blockchain provides transparency, it can also expose sensitive information if not properly managed.

- **Solutions:**
    - **Privacy-Preserving Techniques:** Implementing techniques such as differential privacy and federated learning can help protect individual privacy while still enabling data analysis.
    - **Zero-Knowledge Proofs:** Using zero-knowledge proofs in blockchain systems can allow for the verification of transactions without revealing sensitive information.
    - **Strong Data Governance:** Establishing robust data governance frameworks can ensure that data is collected, stored, and processed in a way that respects privacy rights.

7.4.2 Regulatory and Policy Issues

Regulatory Frameworks: Developing comprehensive regulatory frameworks is essential to govern the ethical use of AI, quantum computing, and blockchain.

- **Challenges:**
    - **Rapid Technological Change:** The fast pace of technological advancement can outstrip the ability of regulatory bodies to develop appropriate frameworks.

- **Complexity:** The complexity and interdependence of these technologies make it difficult to create effective and comprehensive regulations.

- **Solutions:**

    - **Adaptive Regulation:** Developing adaptive regulatory frameworks that can evolve with technological advancements is crucial. This can involve creating flexible guidelines and principles rather than rigid rules.

    - **Stakeholder Collaboration:** Engaging stakeholders from industry, academia, and civil society in the regulatory process can ensure that diverse perspectives are considered.

    - **Sandbox Environments:** Regulatory sandboxes can allow for the testing of new technologies in a controlled environment, providing insights that can inform the development of regulations.

International Cooperation: International cooperation is vital for creating standards and policies for emerging technologies.

- **Challenges:**

    - **Divergent National Interests:** Different countries may have varying priorities and interests, making it challenging to reach consensus on international standards.

    - **Regulatory Fragmentation:** Inconsistent regulations across countries can create barriers to innovation and adoption.

- **Solutions:**

    - **International Organizations:** Leveraging international organizations, such as the United

Nations and the International Telecommunication Union, to facilitate cooperation and harmonize standards.

- **Bilateral and Multilateral Agreements:** Establishing bilateral and multilateral agreements can help align national regulations and promote cross-border collaboration.
- **Global Forums:** Participating in global forums and conferences can foster dialogue and cooperation among stakeholders from different countries.

7.4.3 Security Risks

Quantum Security: Quantum computing poses significant security risks to current cryptographic systems, necessitating the development of quantum-resistant cryptography.

- **Challenges:**
  - **Cryptographic Vulnerability:** Quantum computers have the potential to break widely used cryptographic algorithms, compromising data security.
  - **Transition Period:** The transition to quantum-resistant cryptography must be managed carefully to avoid vulnerabilities during the interim period.

- **Solutions:**
  - **Quantum-Resistant Algorithms:** Developing and implementing quantum-resistant cryptographic algorithms that can withstand attacks from quantum computers.
  - **Proactive Transition:** Proactively planning and executing the transition to quantum-resistant cryptography, including updating standards and protocols.

- **Collaboration:** Collaboration between academia, industry, and government to accelerate research and development in quantum cryptography.

Blockchain Vulnerabilities: Identifying and mitigating vulnerabilities in blockchain systems is essential to ensure their robustness and security.

- **Challenges:**
  - **51% Attacks:** Blockchain networks are vulnerable to 51% attacks, where a malicious actor gains control of the majority of the network's computational power.
  - **Smart Contract Bugs:** Vulnerabilities in smart contracts can be exploited, leading to financial losses and security breaches.

- **Solutions:**
  - **Network Decentralization:** Ensuring sufficient decentralization of blockchain networks to make 51% attacks more difficult and less likely.
  - **Formal Verification:** Using formal verification methods to mathematically prove the correctness of smart contracts and identify potential vulnerabilities.
  - **Regular Audits:** Conducting regular security audits of blockchain systems and smart contracts to detect and address vulnerabilities.

7.4.4 Societal Adaptation

Digital Literacy: Enhancing digital literacy is crucial to prepare society for the widespread adoption of advanced technologies.

- **Challenges:**
  - **Knowledge Gap:** Many people lack the necessary skills and knowledge to understand and effectively use advanced technologies.

- **Access to Education:** Unequal access to education and training opportunities can exacerbate social inequalities.

- **Solutions:**

    - **Educational Programs:** Developing and implementing educational programs that focus on digital literacy and the basics of AI, quantum computing, and blockchain.

    - **Public Awareness Campaigns:** Conducting public awareness campaigns to inform people about the potential benefits and risks of these technologies.

    - **Inclusive Education:** Ensuring that educational opportunities are accessible to all, regardless of socioeconomic status or geographic location.

Social Equity: Ensuring equitable access to technological advancements is essential to prevent widening social and economic inequalities.

- **Challenges:**

    - **Digital Divide:** The digital divide can lead to unequal access to the benefits of advanced technologies, exacerbating existing inequalities.

    - **Economic Displacement:** Technological advancements can displace workers in certain industries, leading to economic instability and inequality.

- **Solutions:**

    - **Universal Access:** Promoting universal access to advanced technologies through initiatives such as affordable internet access and digital infrastructure development.

- **Reskilling and Upskilling:** Implementing programs to reskill and upskill workers, preparing them for new roles in an AI-driven economy.
- **Social Safety Nets:** Strengthening social safety nets to support individuals and communities affected by technological displacement, ensuring that they have the resources to adapt and thrive.

Conclusion: Addressing the potential challenges posed by the convergence of quantum computing, AI, and blockchain is critical to ensuring that these technologies are developed and deployed responsibly. By focusing on ethical dilemmas, regulatory and policy issues, security risks, and societal adaptation, we can navigate the complexities of technological advancement and create a more equitable, secure, and prosperous future.

This comprehensive exploration of potential challenges and solutions highlights the importance of proactive planning and collaboration in harnessing the full potential of quantum computing, AI, and blockchain technologies. By examining specific challenges and proposing actionable solutions, readers can appreciate the steps needed to navigate the ethical, regulatory, security, and societal complexities associated with these transformative technologies.

7.5 Preparing for a Quantum, AI, and Blockchain Future

As we stand on the brink of a technological revolution driven by quantum computing, artificial intelligence (AI), and blockchain, it is crucial to prepare strategically for the changes ahead. This section outlines strategic roadmaps for businesses, government initiatives, education and training approaches, and public awareness and engagement strategies to navigate the future effectively.

7.5.1 Strategic Roadmaps

Roadmaps for Businesses: To stay competitive and harness the full potential of emerging technologies, businesses need to develop comprehensive strategies that integrate quantum computing, AI, and blockchain.

- **Adoption Strategies:**
    - **Assessment and Planning:** Businesses should assess the potential impact of these technologies on their operations and develop a strategic plan for adoption. This includes identifying key areas where technology can add value, such as enhancing product offerings, optimizing processes, and improving customer experiences.
    - **Investment in R&D:** Investing in research and development is essential to stay ahead of the curve. Businesses should allocate resources to explore innovative applications of quantum computing, AI, and blockchain, fostering a culture of continuous innovation.
    - **Partnerships and Collaboration:** Forming strategic partnerships with technology providers, startups, and academic institutions can accelerate the adoption of these technologies. Collaborations can provide access to cutting-edge research, talent, and resources.

- **Implementation Frameworks:**
    - **Pilot Projects:** Implementing pilot projects allows businesses to test the feasibility and impact of new technologies on a small scale before full-scale deployment. This approach helps mitigate risks and identify best practices.
    - **Scalability Planning:** Businesses should plan for scalability from the outset, ensuring that their

infrastructure and processes can support the widespread adoption of new technologies.
- **Change Management:** Managing change effectively is crucial for successful technology adoption. This includes training employees, communicating the benefits of new technologies, and addressing any resistance to change.

Government Initiatives: Governments play a pivotal role in fostering innovation and managing the technological transition. Implementing supportive policies and programs can drive the adoption of quantum computing, AI, and blockchain.

- **Innovation Policies:**
  - **Research Funding:** Governments should allocate funding to support research in quantum computing, AI, and blockchain. Grants and subsidies can encourage academic and industry research, accelerating technological advancements.
  - **Regulatory Frameworks:** Developing adaptive regulatory frameworks that balance innovation with ethical considerations and security is essential. This includes setting standards for data privacy, AI ethics, and blockchain security.
  - **Tax Incentives:** Providing tax incentives for businesses that invest in emerging technologies can stimulate adoption and innovation.

- **Public Sector Adoption:**
  - **Government Services:** Implementing quantum computing, AI, and blockchain in public services can enhance efficiency, transparency, and citizen engagement. Examples include using AI for public health analytics, blockchain for secure voting

systems, and quantum computing for complex policy modeling.

- **Digital Infrastructure:** Investing in digital infrastructure, such as high-speed internet and cloud computing capabilities, can support the deployment of advanced technologies across the public and private sectors.

7.5.2 Education and Training

Curriculum Development: To build a skilled workforce capable of advancing and applying new technologies, educational institutions must incorporate quantum computing, AI, and blockchain into their curricula.

- **Interdisciplinary Programs:** Developing interdisciplinary programs that combine computer science, engineering, mathematics, and ethics can provide students with a comprehensive understanding of these technologies.

    - **Example:** Universities can offer specialized degrees or certificates in quantum computing, AI, and blockchain, with courses covering theoretical foundations, practical applications, and ethical considerations.

- **Hands-On Training:**

    - **Laboratories and Workshops:** Establishing laboratories and workshops equipped with state-of-the-art technology can provide students with hands-on experience. Collaborations with industry can offer access to real-world projects and mentorship.

    - **Internships and Co-ops:** Partnering with businesses to offer internships and cooperative education programs can help students gain practical experience and build industry connections.

Lifelong Learning: Promoting lifelong learning and continuous upskilling is essential to keep pace with technological advancements.

- **Online Courses and MOOCs:** Offering online courses and massive open online courses (MOOCs) on quantum computing, AI, and blockchain can make education accessible to a broader audience.
  - **Example:** Platforms like Coursera, edX, and Udacity can provide specialized courses and certifications that professionals can pursue alongside their careers.
- **Corporate Training Programs:** Businesses should invest in continuous training programs to keep their workforce updated with the latest technological developments.
  - **Example:** Companies can partner with educational institutions or online platforms to offer customized training programs for their employees.

7.5.3 Public Awareness and Engagement

Community Involvement: Engaging communities in discussions about the future of technology and its societal impacts is crucial for fostering understanding and acceptance.

- **Public Forums and Workshops:** Organizing public forums, workshops, and seminars can provide platforms for dialogue between technologists, policymakers, and the public. These events can address concerns, dispel myths, and highlight the benefits of emerging technologies.
  - **Example:** Local governments and community organizations can host town hall meetings and hackathons to involve citizens in technological discussions and innovations.
- **Citizen Science Projects:** Encouraging citizen science projects can engage the public in hands-on technological exploration and innovation.

- **Example:** Initiatives like citizen-driven environmental monitoring using blockchain for data transparency or AI-driven community health projects.

Transparency and Dialogue: Ensuring transparency in technological development and fostering open dialogue between developers, policymakers, and the public is essential.

- **Ethical Review Boards:** Establishing ethical review boards can oversee the development and deployment of emerging technologies, ensuring they align with societal values and ethical standards.
    - **Example:** Boards can include representatives from academia, industry, government, and civil society to provide diverse perspectives.
- **Public Communication:** Clear and transparent communication about the goals, progress, and implications of technological projects can build trust and understanding.
    - **Example:** Governments and businesses can use websites, social media, and public reports to keep the public informed and involved in decision-making processes.
- **MOTMSDD Approach:** Leveraging the Metaverse of the Minds Social Direct Democracy (MOTMSDD) approach can enhance public engagement and participatory democracy.
    - **Implementation:** Using AI and blockchain to create a digital metaverse where citizens can participate directly in policy-making processes, ensuring their voices are heard and their needs are addressed.
    - **Impact:** This approach can foster a more inclusive and democratic society, where public decision-making is transparent, accountable, and reflective of the collective will.

Conclusion: Preparing for a future shaped by quantum computing, AI, and blockchain requires strategic planning, education, public engagement, and collaboration across sectors. By developing strategic roadmaps, enhancing education and training, and fostering public awareness and engagement, we can navigate the complexities of technological advancement and create a future that benefits all of society.

This comprehensive exploration of preparing for a quantum, AI, and blockchain future highlights the importance of proactive strategies, education, and public engagement. By examining specific actions and their implications, readers can appreciate the steps needed to navigate the challenges and opportunities associated with these transformative technologies.

7.6 Conclusion: Shaping the Future

As we stand on the cusp of a technological revolution, the convergence of quantum computing, artificial intelligence (AI), and blockchain technology offers a transformative potential that will reshape industries, economies, and everyday life. This chapter concludes by envisioning the future shaped by these advancements and calling for collaborative efforts to ensure these technologies benefit society as a whole.

7.6.1 The Promise of Convergence

Transformative Potential: The convergence of quantum computing, AI, and blockchain holds the promise of unprecedented advancements across multiple sectors.

- **Healthcare:** AI's predictive analytics, quantum computing's processing power, and blockchain's secure data management will revolutionize medical diagnostics, personalized medicine, and health records management. These

technologies will enable earlier disease detection, tailored treatments, and secure patient data sharing, ultimately leading to improved patient outcomes and healthcare efficiency.
- **Finance:** The integration of AI and quantum computing in financial modeling, risk assessment, and fraud detection will create more robust and resilient financial systems. Blockchain technology will facilitate decentralized finance (DeFi) platforms, enabling secure, transparent, and efficient financial transactions without traditional intermediaries.
- **Urban Development:** AI-powered smart cities will optimize resource allocation, traffic management, and urban planning, enhancing sustainability and quality of life. Quantum simulations will enable more accurate and comprehensive urban planning, while blockchain governance systems will ensure transparent and participatory decision-making processes.
- **Education:** Personalized learning experiences powered by AI will adapt to individual student needs, making education more effective and engaging. Quantum computing education programs will prepare the workforce for future technological demands, ensuring a continuous supply of skilled professionals.
- **Public Policy:** AI will support data-driven policy analysis and decision-making, ensuring policies are informed by accurate and comprehensive data. Blockchain technology will enhance transparency and reduce corruption in governance, while the MOTMSDD approach will enable direct citizen participation in public decision-making.

Vision for the Future: The convergence of these technologies will drive innovation, prosperity, and well-being, shaping a future where:

- **Innovation:** Continuous technological advancements will lead to the development of new products, services, and solutions that address global challenges and improve quality of life.

- **Prosperity:** Economic growth will be fueled by the widespread adoption of advanced technologies, creating new industries, job opportunities, and wealth.
- **Well-being:** Enhanced healthcare, sustainable urban environments, and efficient governance will contribute to the overall well-being of individuals and communities.

7.6.2 Call to Action

Collaboration and Innovation: To fully realize the potential of these technologies, collaboration across sectors is essential.

- **Cross-Sector Collaboration:** Governments, businesses, academic institutions, and civil society must work together to develop and implement innovative solutions. Collaborative research initiatives, public-private partnerships, and international cooperation will accelerate technological advancements and ensure they are used responsibly.
- **Ethical Innovation:** Innovators must prioritize ethical considerations, ensuring that technologies are developed and deployed in ways that promote fairness, transparency, and accountability. Ethical guidelines, regulatory frameworks, and industry standards will play a crucial role in guiding responsible innovation.

Empowering Society: Empowering society to harness the benefits of these technologies while addressing their challenges is paramount.

- **Education and Training:** Comprehensive education and training programs are essential to prepare individuals for the jobs of the future and ensure they have the skills needed to thrive in a rapidly evolving technological landscape. Lifelong learning and continuous upskilling will be crucial to keep pace with advancements.
- **Public Engagement:** Engaging the public in discussions about the future of technology and its societal impacts will foster understanding, trust, and acceptance. Transparent communication, public forums, and community involvement

will ensure that technological developments align with societal values and needs.
- **Equitable Access:** Ensuring equitable access to advanced technologies is vital to prevent widening social and economic inequalities. Policies and initiatives that promote digital inclusion, affordable access, and reskilling programs will help bridge the digital divide and ensure that all members of society can benefit from technological advancements.

Conclusion: The convergence of quantum computing, AI, and blockchain technology represents a monumental opportunity to reshape our world for the better. By embracing collaboration, ethical innovation, and societal empowerment, we can harness the transformative potential of these technologies to create a future that is prosperous, equitable, and sustainable. Let us embark on this journey with a shared vision and a commitment to shaping a brighter future for all of humanity.

This comprehensive conclusion highlights the transformative potential of integrating quantum computing, AI, and blockchain technologies and emphasizes the importance of collaboration, ethical innovation, and societal empowerment. By outlining a visionary outlook and a call to action, readers are encouraged to actively participate in shaping a future that leverages these technologies for the benefit of mankind.

- .

Chapter 8: Practical Guide for Businesses and Individuals

8.1 Preparing Businesses for Emerging Technologies

As the technological landscape evolves with the integration of quantum computing, artificial intelligence (AI), and blockchain, businesses must strategically prepare to harness these advancements. This section provides a detailed roadmap for businesses to understand, plan, and implement these cutting-edge technologies to stay competitive and drive innovation.

8.1.1 Understanding the Technology Landscape

Comprehensive Overview: Businesses need a thorough understanding of the capabilities and potential of AI, quantum computing, and blockchain.

- **Artificial Intelligence:**
    - **Current Capabilities:** AI systems can process vast amounts of data, identify patterns, make predictions, and automate tasks. Key areas include machine learning, natural language processing, and computer vision.
    - **Future Potential:** Advancements in AI will lead to more sophisticated systems capable of understanding and interacting with humans seamlessly, enhancing decision-making, and driving innovation across sectors.

- **Quantum Computing:**

- **Current Capabilities:** Quantum computers leverage quantum bits (qubits) to perform complex calculations at unprecedented speeds. They excel in optimization problems, cryptographic analysis, and molecular simulations.
- **Future Potential:** Quantum computing is expected to revolutionize fields such as drug discovery, financial modeling, and artificial intelligence, providing solutions to problems currently deemed intractable.

- **Blockchain:**
  - **Current Capabilities:** Blockchain provides a secure, decentralized ledger for recording transactions. It ensures transparency, immutability, and security, making it ideal for applications in finance, supply chain management, and digital identity.
  - **Future Potential:** The widespread adoption of blockchain could lead to the development of decentralized autonomous organizations (DAOs), transforming governance, finance, and digital interaction.

Industry-Specific Applications: Understanding how these technologies can be applied across various industries is crucial.

- **Healthcare:**

- **AI:** Predictive analytics for disease diagnosis, personalized medicine, and AI-driven clinical trials.
- **Quantum Computing:** Accelerated drug discovery and personalized treatment planning.
- **Blockchain:** Secure patient records management and transparent drug supply chains.

- **Finance:**
    - **AI:** Algorithmic trading, risk management, and fraud detection.
    - **Quantum Computing:** Enhanced financial modeling and optimization.
    - **Blockchain:** Decentralized finance (DeFi) platforms and secure cross-border transactions.

- **Retail:**
    - **AI:** Personalized marketing, inventory management, and customer service automation.
    - **Quantum Computing:** Optimization of supply chains and pricing strategies.
    - **Blockchain:** Transparent supply chains and authentication of products.

- **Manufacturing:**
    - **AI:** Predictive maintenance, quality control, and process automation.
    - **Quantum Computing:** Material science research and optimization of manufacturing processes.

- **Blockchain:** Traceability of products and secure transactions.

8.1.2 Strategic Planning and Road Mapping

Technology Roadmap Development: Creating a technology roadmap that aligns with business goals and industry trends is essential.

- **Steps for Development:**
    - **Assessment:** Evaluate the current technological landscape and identify relevant trends.
    - **Alignment:** Ensure the roadmap aligns with the business's strategic goals and objectives.
    - **Phased Implementation:** Develop a phased plan for integrating new technologies, starting with pilot projects.

Investment Planning: Budgeting and investing in emerging technologies is critical for successful adoption.

- **Guidance:**
    - **Prioritization:** Identify and prioritize high-impact projects that offer the most significant benefits.
    - **Resource Allocation:** Allocate resources effectively, ensuring sufficient funding for R&D, training, and infrastructure.
    - **Risk Management:** Incorporate risk management strategies to address potential challenges and uncertainties.

8.1.3 Building a Tech-Savvy Workforce

Skills Assessment: Assessing the current skills of the workforce and identifying gaps is the first step in building a tech-savvy team.

- **Methods:**
    - **Surveys and Assessments:** Conduct surveys and skills assessments to understand the current capabilities of employees.
    - **Gap Analysis:** Identify skill gaps that need to be addressed to meet future technological demands.

Training and Development Programs: Creating training programs and continuous learning opportunities is crucial.

- **Programs:**
    - **Upskilling and Reskilling:** Develop programs to upskill current employees in emerging technologies.
    - **Continuous Learning:** Encourage a culture of continuous learning through workshops, online courses, and certifications.
    - **Partnerships:** Partner with educational institutions and online platforms to provide comprehensive training resources.

8.1.4 Partnerships and Collaborations

Strategic Alliances: Forming partnerships with tech companies, research institutions, and startups can leverage external expertise and resources.

- **Strategies:**
  - **Tech Companies:** Collaborate with leading technology providers for access to the latest advancements.
  - **Research Institutions:** Engage with academic institutions for cutting-edge research and development.
  - **Startups:** Partner with innovative startups to foster a culture of innovation and agility.

Industry Consortia: Joining industry consortia and collaborative platforms keeps businesses updated on technological advancements and best practices.

- **Benefits:**
  - **Knowledge Sharing:** Gain insights from industry peers and participate in collaborative research.
  - **Standardization:** Contribute to the development of industry standards and protocols.
  - **Networking:** Build a network of industry experts and innovators.

8.1.5 Innovation and R&D

Fostering Innovation: Encouraging a culture of innovation within the organization is essential for staying competitive.

- **Strategies:**
  - **Innovation Hubs:** Establish innovation hubs or labs to experiment with new ideas and technologies.

- **Incentives:** Create incentives for employees to innovate and contribute ideas.
- **Collaboration:** Foster cross-departmental collaboration to drive creative solutions.

R&D Investment: Investing in research and development is crucial for staying ahead of the technological curve.

- **Importance:**
    - **Competitive Edge:** Maintain a competitive edge by continually exploring new technologies and applications.
    - **Product Development:** Accelerate the development of innovative products and services.
    - **Market Leadership:** Position the company as a leader in technological advancements.

8.1.6 Integrating New Technologies

Pilot Projects: Launching pilot projects to test the feasibility and impact of new technologies before full-scale implementation is vital.

- **Implementation:**
    - **Small Scale:** Start with small-scale projects to minimize risk and gather insights.
    - **Evaluation:** Assess the outcomes of pilot projects to determine their viability.
    - **Scaling Up:** Use the insights gained to scale up successful projects across the organization.

Scalability Considerations: Ensuring that technology solutions are scalable and can grow with the business is crucial.

- **Strategies:**
    - **Modular Design:** Design solutions with modular components that can be easily expanded.
    - **Infrastructure:** Invest in scalable infrastructure that can support future growth.
    - **Future-Proofing:** Plan for future technological advancements and ensure compatibility with evolving standards.

Conclusion: Preparing businesses for emerging technologies requires a comprehensive approach that includes understanding the technology landscape, strategic planning, building a tech-savvy workforce, fostering innovation, and ensuring scalability. By following these guidelines, businesses can effectively leverage quantum computing, AI, and blockchain to drive growth, innovation, and competitive advantage.

This detailed roadmap provides businesses with the necessary strategies and steps to prepare for the integration of quantum computing, AI, and blockchain technologies. By addressing various aspects of preparation, from understanding the technologies to fostering innovation and scalability, readers can gain a comprehensive understanding of how to navigate the technological future.

8.2 Leveraging AI for Business Success

Artificial intelligence (AI) has become a critical tool for driving business success in the modern digital landscape. By developing a robust AI strategy, managing data effectively, implementing AI tools, and continuously monitoring and optimizing AI initiatives, businesses can unlock significant value and maintain a competitive edge. This section provides a detailed guide for leveraging AI to achieve business success.

8.2.1 AI Strategy Development

Defining AI Objectives: Setting clear objectives for AI initiatives is crucial to ensure they align with the overall business goals and deliver measurable value.

- **Objective Setting:**
    - **Business Alignment:** Ensure that AI objectives are aligned with the broader strategic goals of the business. This alignment helps in prioritizing AI initiatives that directly contribute to business growth and efficiency.
    - **Specific, Measurable, Achievable, Relevant, Time-bound (SMART) Goals:** Define SMART goals for AI projects to provide clear direction and benchmarks for success.
    - **Stakeholder Engagement:** Involve key stakeholders from different departments in the objective-setting process to ensure comprehensive understanding and buy-in.

AI Use Cases: Identifying high-impact use cases for AI in the business helps in focusing efforts on areas with the greatest potential for value creation.

- **Customer Service Automation:**

    - **Chatbots and Virtual Assistants:** Implement AI-driven chatbots and virtual assistants to handle customer inquiries, provide support, and enhance customer experience.
    - **Case Study:** Analyse the implementation of AI chatbots in a customer service center, highlighting improvements in response times and customer satisfaction.

- **Predictive Analytics:**

    - **Demand Forecasting:** Use AI algorithms to analyse historical data and predict future demand, helping in inventory management and production planning.
    - **Risk Management:** Apply predictive analytics to identify and mitigate risks, enhancing decision-making processes.

- **Personalized Marketing:**

    - **Targeted Campaigns:** Utilize AI to analyse customer data and deliver personalized marketing messages, increasing engagement and conversion rates.
    - **Customer Segmentation:** Implement AI-driven segmentation to better understand customer

behavior and preferences, tailoring marketing strategies accordingly.

8.2.2 Data Management

Data Collection and Storage: Effective data management is the foundation of successful AI initiatives. Best practices for collecting, storing, and managing data are essential to fuel AI applications.

- **Data Collection:**
    - **Data Sources:** Identify and integrate various data sources, including internal databases, external APIs, and third-party data providers.
    - **Real-Time Data:** Implement real-time data collection mechanisms to provide up-to-date information for AI models.

- **Data Storage:**
    - **Cloud Solutions:** Utilize cloud storage solutions for scalable and secure data storage.
    - **Data Lakes and Warehouses:** Implement data lakes and warehouses to store structured and unstructured data, providing a central repository for AI applications.

Data Quality and Governance: Ensuring data quality and establishing governance frameworks are critical for managing data ethically and responsibly.

- **Data Quality:**

- **Data Cleaning:** Implement data cleaning processes to remove duplicates, correct errors, and fill missing values.
- **Validation:** Use automated validation techniques to ensure data accuracy and consistency.

- **Data Governance:**
  - **Ethical Standards:** Establish ethical standards for data usage, ensuring compliance with privacy regulations such as GDPR and CCPA.
  - **Governance Frameworks:** Develop governance frameworks to manage data access, security, and usage policies.

8.2.3 AI Implementation

Choosing AI Tools: Selecting the right AI tools and platforms that meet the specific needs of the business is crucial for successful implementation.

- **Tool Selection:**
  - **Platform Evaluation:** Evaluate AI platforms based on features, scalability, integration capabilities, and cost.
  - **Vendor Assessment:** Assess vendors for their expertise, support services, and track record in delivering AI solutions.

Integrating AI Systems: Integrating AI systems with existing IT infrastructure and business processes ensures seamless operation and maximizes the value of AI initiatives.

- **System Integration:**
    - **API Integration:** Use APIs to integrate AI tools with existing software systems, enabling data flow and process automation.
    - **Workflow Automation:** Implement AI-driven automation in business workflows to enhance efficiency and reduce manual intervention.

8.2.4 Monitoring and Optimization

Performance Metrics: Defining key performance indicators (KPIs) to monitor the success of AI initiatives is essential for assessing their impact and guiding continuous improvement.

- **KPIs:**
    - **Accuracy:** Measure the accuracy of AI models in predicting outcomes or classifying data.
    - **Efficiency:** Assess improvements in process efficiency, such as reduced processing time or increased throughput.
    - **Customer Satisfaction:** Evaluate customer satisfaction metrics to gauge the impact of AI-driven customer service solutions.

Continuous Improvement: Implementing feedback loops to continuously improve AI models and applications ensures they remain effective and relevant.

- **Feedback Mechanisms:**
    - **User Feedback:** Collect feedback from users to identify areas for improvement and refine AI models accordingly.

- **Model Retraining:** Regularly retrain AI models with new data to maintain accuracy and adapt to changing conditions.
- **Performance Reviews:** Conduct periodic performance reviews to assess the effectiveness of AI initiatives and make necessary adjustments.

Conclusion: Leveraging AI for business success requires a comprehensive approach that encompasses strategy development, data management, implementation, and continuous optimization. By setting clear objectives, managing data effectively, choosing the right tools, and continuously monitoring and improving AI initiatives, businesses can harness the full potential of AI to drive growth, innovation, and competitive advantage.

This detailed guide provides businesses with the necessary steps to leverage AI effectively. By addressing various aspects of AI integration, from strategy development to continuous optimization, readers can gain a comprehensive understanding of how to implement AI initiatives that align with business goals and deliver measurable value.

8.3 Harnessing the Power of Quantum Computing

Quantum computing promises to revolutionize industries with its unparalleled processing power and ability to solve complex problems beyond the reach of classical computers. Businesses must strategically prepare to harness this power by assessing their readiness, building expertise, and implementing pilot projects. This section provides a comprehensive guide for businesses to effectively integrate quantum computing into their operations.

8.3.1 Quantum Readiness Assessment

Current Capabilities: Evaluating the current capabilities of the business in relation to quantum computing is the first step towards integration.

- **Infrastructure Assessment:**
    - **Existing IT Infrastructure:** Assess the existing IT infrastructure to determine if it can support quantum computing initiatives, including compatibility with quantum software and hardware requirements.
    - **Data Requirements:** Evaluate the data processing needs and storage capabilities to ensure they align with the demands of quantum computing.

- **Skill Set Evaluation:**
    - **Current Workforce Skills:** Assess the current skill set of the workforce to identify gaps in quantum computing knowledge and expertise.
    - **Training Needs:** Determine the training needs for existing employees to equip them with the necessary skills to support quantum initiatives.

Identifying Opportunities: Identifying business processes and challenges that could benefit from quantum computing helps in prioritizing potential use cases.

- **Optimization Problems:**
    - **Supply Chain Optimization:** Use quantum algorithms to optimize supply chain logistics, reducing costs and improving efficiency.
    - **Financial Modeling:** Apply quantum computing to complex financial models for enhanced risk assessment and portfolio optimization.

- **Research and Development:**
    - **Drug Discovery:** Leverage quantum computing for faster and more accurate drug discovery by simulating molecular interactions.
    - **Material Science:** Utilize quantum simulations to develop new materials with desired properties for various industrial applications.

8.3.2 Quantum Literacy

Employee Training: Providing training programs to educate employees about the basics and potential of quantum computing is essential for building a knowledgeable workforce.

- **Training Programs:**
    - **Introductory Courses:** Offer introductory courses on quantum computing principles, covering topics such as qubits, entanglement, and quantum gates.
    - **Advanced Training:** Provide advanced training on quantum algorithms, programming languages (e.g., Qiskit, Cirq), and practical applications.

- **Certifications:**
    - **Professional Certifications:** Encourage employees to pursue professional certifications in quantum computing to validate their skills and knowledge.

Awareness Campaigns: Running awareness campaigns to keep the workforce informed about the latest developments in quantum technology fosters a culture of continuous learning and innovation.

- **Internal Communications:**
    - **Newsletters and Updates:** Distribute regular newsletters and updates on quantum computing advancements and their potential impact on the business.
    - **Workshops and Seminars:** Organize workshops and seminars featuring industry experts to discuss recent trends and breakthroughs in quantum technology.
- **Knowledge Sharing:**
    - **Internal Forums:** Create internal forums and discussion groups where employees can share knowledge, ask questions, and collaborate on quantum computing projects.

8.3.3 Building Quantum Expertise

Hiring Specialists: Hiring quantum computing specialists and building an in-house team with expertise in quantum algorithms and applications is crucial for developing and implementing quantum initiatives.

- **Talent Acquisition:**
  - **Job Descriptions:** Develop detailed job descriptions outlining the required skills and qualifications for quantum computing roles.
  - **Recruitment Strategies:** Use targeted recruitment strategies, including partnerships with universities and participation in quantum computing conferences, to attract top talent.

- **Team Development:**
  - **Interdisciplinary Teams:** Form interdisciplinary teams that combine quantum expertise with domain-specific knowledge to address business challenges effectively.

External Consultants: Engaging with external consultants and quantum computing firms provides access to specialized knowledge and skills that may not be available in-house.

- **Consulting Partnerships:**
  - **Engagement Models:** Establish engagement models with external consultants, including short-term projects, long-term collaborations, and advisory roles.
  - **Knowledge Transfer:** Ensure that consultants transfer knowledge to the internal team through training sessions and documentation.

- **Vendor Selection:**
    - **Evaluation Criteria:** Evaluate quantum computing firms based on their expertise, track record, and ability to deliver customized solutions.
    - **Collaboration Agreements:** Develop collaboration agreements that outline the scope, objectives, and deliverables of the engagement.

8.3.4 Quantum Pilot Projects

Proof of Concept: Developing proof-of-concept projects to explore the practical applications of quantum computing within the business allows for experimentation and learning.

- **Project Selection:**
    - **High-Impact Areas:** Select pilot projects in high-impact areas where quantum computing can demonstrate significant benefits.
    - **Feasibility Studies:** Conduct feasibility studies to assess the technical and operational viability of the proposed projects.

- **Implementation:**
    - **Prototype Development:** Develop prototypes to test quantum algorithms and solutions in real-world scenarios.
    - **Performance Evaluation:** Evaluate the performance of the prototypes based on predefined metrics and criteria.

Scalability Testing: Testing the scalability and real-world impact of quantum solutions before broader implementation ensures that they can handle increased workloads and deliver consistent results.

- **Scalability Assessments:**
    - **Load Testing:** Perform load testing to determine the capacity and performance limits of quantum solutions.
    - **Stress Testing:** Conduct stress testing to identify potential bottlenecks and areas for improvement.

- **Deployment Planning:**
    - **Pilot Expansion:** Gradually expand successful pilot projects to additional business units or processes.
    - **Full-Scale Implementation:** Develop a detailed plan for full-scale implementation, including timelines, resource allocation, and risk management strategies.

Conclusion: Harnessing the power of quantum computing requires a comprehensive approach that includes assessing readiness, building expertise, and implementing pilot projects. By strategically integrating quantum computing into their operations, businesses can unlock new opportunities, drive innovation, and maintain a competitive edge in the rapidly evolving technological landscape.

This detailed guide provides businesses with the necessary steps to harness the power of quantum computing effectively. By addressing various aspects of preparation, from readiness assessment to scalability testing, readers can gain a comprehensive understanding of how to integrate quantum computing initiatives that align with business goals and deliver measurable value.

8.4 Implementing Blockchain Solutions

Implementing blockchain solutions can transform various aspects of business operations by enhancing transparency, security, and efficiency. This section provides a comprehensive guide on conducting feasibility studies, developing blockchain applications, ensuring security and compliance, and integrating blockchain with existing systems.

8.4.1 Blockchain Feasibility Study

Assessing Use Cases: Conducting a feasibility study is the first step in identifying viable use cases for blockchain within the business.

- **Use Case Identification:**
    - **Business Processes:** Evaluate current business processes to identify areas where blockchain can add value, such as supply chain management, financial transactions, and data sharing.
    - **Pain Points:** Identify pain points that blockchain technology can address, such as inefficiencies, lack of transparency, and security vulnerabilities.
- **Stakeholder Involvement:**

- **Interviews and Workshops:** Conduct interviews and workshops with key stakeholders to gather insights and understand their needs and expectations.
- **Cross-Functional Teams:** Form cross-functional teams to ensure a comprehensive assessment of potential use cases from different perspectives.

Cost-Benefit Analysis: Performing a cost-benefit analysis helps evaluate the potential return on investment for blockchain projects.

- **Cost Estimation:**
    - **Development Costs:** Estimate the costs associated with developing and implementing blockchain solutions, including software, hardware, and personnel.
    - **Operational Costs:** Consider ongoing operational costs, such as maintenance, updates, and transaction fees.

- **Benefit Assessment:**
    - **Efficiency Gains:** Quantify the efficiency gains from automating and streamlining processes with blockchain.
    - **Risk Reduction:** Evaluate the potential reduction in risks, such as fraud and data breaches, due to enhanced security features.
    - **Revenue Opportunities:** Identify new revenue opportunities enabled by blockchain, such as new business models or services.

8.4.2 Blockchain Development

Choosing the Right Platform: Selecting the appropriate blockchain platform is crucial for the success of blockchain initiatives.

- **Platform Evaluation:**

- **Functionality:** Assess the functionality and features of various blockchain platforms, such as Ethereum, Hyperledger, and Corda.
- **Scalability:** Consider the scalability of the platform to ensure it can handle the expected transaction volume.
- **Community and Support:** Evaluate the strength of the developer community and the availability of support and resources.

Smart Contract Development: Developing and deploying smart contracts automates and secures business processes.

- **Smart Contract Design:**
  - **Business Logic:** Define the business logic and rules that the smart contract will enforce.
  - **Security Considerations:** Incorporate security measures to prevent vulnerabilities and ensure the integrity of the smart contract.
- **Development Tools:**
  - **Integrated Development Environments (IDEs):** Use IDEs like Remix or Truffle for smart contract development.
  - **Testing Frameworks:** Implement testing frameworks to thoroughly test smart contracts before deployment.
- **Deployment:**
  - **Deployment Strategy:** Plan the deployment strategy, including the choice of public or private blockchain and the deployment process.

- **Monitoring and Maintenance:** Establish monitoring and maintenance procedures to ensure the smooth operation of smart contracts.

8.4.3 Ensuring Security and Compliance

Security Best Practices: Implementing security best practices is essential to protect blockchain applications from vulnerabilities and attacks.

- **Security Measures:**
    - **Encryption:** Use strong encryption techniques to protect data on the blockchain.
    - **Access Control:** Implement robust access control mechanisms to restrict unauthorized access to blockchain applications.
    - **Regular Audits:** Conduct regular security audits to identify and address vulnerabilities.

- **Incident Response:**
    - **Response Plan:** Develop an incident response plan to quickly respond to and mitigate security incidents.
    - **Training:** Train employees on security best practices and incident response procedures.

Regulatory Compliance: Ensuring compliance with relevant regulations and standards governing blockchain use in the industry is critical.

- **Regulatory Landscape:**
    - **Understanding Regulations:** Stay informed about the regulatory landscape and requirements for blockchain use in your industry.

- **Compliance Strategies:** Develop strategies to ensure compliance with regulations such as GDPR, CCPA, and industry-specific standards.

- **Data Privacy:**

  - **Privacy by Design:** Implement privacy-by-design principles to protect user data and ensure compliance with data protection regulations.

  - **Anonymization and Pseudonymization:** Use techniques such as anonymization and pseudonymization to protect sensitive data on the blockchain.

8.4.4 Blockchain Integration

Interoperability: Ensuring interoperability between blockchain solutions and existing IT systems is crucial for seamless integration.

- **Integration Standards:**

  - **APIs:** Use standardized APIs to facilitate communication between blockchain and existing systems.

  - **Middleware Solutions:** Implement middleware solutions to bridge the gap between blockchain and legacy systems.

- **Interoperability Frameworks:**

  - **Cross-Chain Solutions:** Explore cross-chain solutions that enable interoperability between different blockchain networks.

  - **Industry Standards:** Adopt industry standards for interoperability to ensure compatibility with other systems and platforms.

Data Migration: Planning and executing data migration strategies to integrate blockchain with legacy systems ensures a smooth transition.

- **Migration Planning:**
    - **Data Mapping:** Map existing data structures to the blockchain format to ensure consistency.
    - **Phased Approach:** Use a phased approach to migrate data gradually, minimizing disruption to business operations.

- **Data Validation:**
    - **Data Integrity Checks:** Perform data integrity checks to ensure the accuracy and completeness of migrated data.
    - **Testing:** Conduct thorough testing to validate the migration process and ensure the seamless operation of integrated systems.

Conclusion: Implementing blockchain solutions involves conducting feasibility studies, developing applications, ensuring security and compliance, and integrating with existing systems. By following these guidelines, businesses can effectively harness the power of blockchain technology to enhance transparency, security, and efficiency, ultimately driving innovation and competitive advantage.

This detailed guide provides businesses with the necessary steps to implement blockchain solutions effectively. By addressing various aspects of blockchain integration, from feasibility studies to data migration, readers can gain a comprehensive understanding of how to leverage blockchain technology to achieve business goals and deliver measurable value.

8.5 Tips for Individuals to Stay Ahead

In a world where technology evolves at an unprecedented pace, staying ahead requires a proactive approach to learning, networking, and gaining practical experience. This section provides actionable tips for individuals to remain competitive and capitalize on opportunities in AI, quantum computing, and blockchain.

8.5.1 Lifelong Learning

Continuous Education: Continuous education is vital for staying updated with the latest advancements in AI, quantum computing, and blockchain.

- **Importance of Lifelong Learning:**
    - **Adapting to Technological Changes:** Technologies are constantly evolving, making it essential to stay informed and adapt to new developments.
    - **Career Advancement:** Continuous education opens doors to new career opportunities and advancements by equipping individuals with in-demand skills.

Online Courses and Certifications: Enrolling in online courses and certification programs provides flexible and accessible ways to gain expertise in emerging technologies.

- **Reputable Platforms:**
    - **Coursera:** Offers courses from top universities and companies on AI, blockchain, and quantum computing.
    - **edX:** Provides a wide range of courses and certifications from institutions like MIT, Harvard, and IBM.

- **Udacity:** Known for its nanodegree programs in AI and blockchain, offering practical, project-based learning.

- **Certification Programs:**

  - **Certified AI Practitioner:** Validates expertise in AI and machine learning.

  - **Blockchain Certification:** Programs like the Certified Blockchain Professional (CBP) offer in-depth knowledge of blockchain technology.

  - **Quantum Computing Courses:** IBM Quantum and other providers offer foundational and advanced courses in quantum computing.

8.5.2 Networking and Community Engagement

Professional Networks: Joining professional networks and communities helps individuals stay connected and learn from peers and experts.

- **Industry Associations:**

  - **IEEE:** The Institute of Electrical and Electronics Engineers provides networking opportunities and resources across various technological fields.

  - **ACM:** The Association for Computing Machinery offers a platform for computing professionals to connect and collaborate.

- **Online Communities:**

  - **LinkedIn Groups:** Join LinkedIn groups focused on AI, blockchain, and quantum computing to engage with industry professionals.

  - **Reddit and Stack Exchange:** Participate in specialized forums to discuss and seek advice on these technologies.

Conferences and Workshops: Attending conferences, workshops, and webinars provides opportunities to learn from industry experts and stay updated on the latest trends.

- **Major Conferences:**
    - **AI Conferences:** Events like NeurIPS, AAAI Conference, and the AI Summit bring together leading researchers and practitioners in AI.
    - **Blockchain Conferences:** Consensus and Blockchain Expo showcase the latest in blockchain technology.
    - **Quantum Computing Events:** Q2B and the IEEE Quantum Week are key events for those interested in quantum computing.

- **Workshops and Webinars:**
    - **Skill-Building Workshops:** Participate in workshops focusing on specific skills, such as machine learning, smart contract development, or quantum algorithms.
    - **Webinars:** Attend webinars hosted by industry leaders to gain insights into emerging trends and best practices.

8.5.3 Practical Experience

Hands-On Projects: Engaging in hands-on projects and practical applications is essential for building real-world experience and deepening understanding.

- **Project Ideas:**
    - **AI Applications:** Develop machine learning models for real-world problems, such as predictive analytics or image recognition.
    - **Blockchain Solutions:** Create and deploy smart contracts on platforms like Ethereum or Hyperledger.

- **Quantum Computing Experiments:** Use platforms like IBM Quantum Experience to explore quantum algorithms and simulations.

- **Open Source Contributions:**
  - **GitHub:** Contribute to open source projects in AI, blockchain, and quantum computing to gain practical experience and collaborate with the global developer community.

Hackathons and Competitions: Participating in hackathons and competitions challenges individuals to apply their skills, fosters innovation, and offers opportunities for recognition.

- **Hackathons:**
  - **AI and Data Science Hackathons:** Platforms like Kaggle host competitions that challenge participants to solve data science problems using AI and machine learning.
  - **Blockchain Hackathons:** Events like ETHGlobal and Chainlink Hackathon provide opportunities to build innovative blockchain applications.
  - **Quantum Computing Challenges:** Participate in quantum computing competitions organized by IBM Quantum and other platforms to solve complex problems.

- **Competitions:**
  - **Innovation Challenges:** Join competitions that focus on technological innovation, offering prizes and recognition for groundbreaking ideas and solutions.
  - **Academic Competitions:** Participate in academic competitions and symposiums to showcase research and projects.

8.5.4 Personal Development

Soft Skills: Developing soft skills is essential to complement technical expertise and excel in the workplace.

- **Critical Thinking:** Enhance problem-solving abilities by developing critical thinking skills to analyse complex issues and make informed decisions.
- **Adaptability:** Cultivate adaptability to navigate the rapidly changing technological landscape and embrace new challenges.
- **Communication:** Improve communication skills to effectively convey complex technical concepts to non-technical stakeholders and collaborate with diverse teams.

Staying Informed: Regularly reading industry publications, research papers, and news is crucial to stay informed about the latest trends and developments.

- **Industry Publications:**
    - **Journals and Magazines:** Subscribe to journals and magazines that focus on AI, quantum computing, and blockchain, such as MIT Technology Review and IEEE Spectrum.
    - **Blogs and News Sites:** Follow blogs and news sites that provide updates and insights into technological advancements.

- **Research Papers:**
    - **Academic Databases:** Access academic databases like Google Scholar to read the latest research papers and stay informed about cutting-edge developments.
    - **Preprint Repositories:** Explore preprint repositories like arXiv to discover new research before it is published in journals.

Conclusion: Staying ahead in the rapidly evolving fields of AI, quantum computing, and blockchain requires a proactive approach to learning, networking, and gaining practical experience. By embracing lifelong learning, engaging with professional communities, participating in hands-on projects and competitions, and developing soft skills, individuals can remain competitive and capitalize on the opportunities presented by these transformative technologies.

This comprehensive guide provides individuals with practical tips to stay ahead in the fast-paced technological landscape. By addressing various aspects of continuous education, networking, practical experience, and personal development, readers can develop a holistic strategy to remain relevant and excel in the fields of AI, quantum computing, and blockchain.

8.6 Future-Proofing Careers and Businesses

As AI, quantum computing, and blockchain technologies continue to evolve, future-proofing careers and businesses becomes essential to navigate the uncertainties and capitalize on new opportunities. This section provides strategies to adapt to change, build resilience, and consider ethical implications, ensuring long-term success and societal benefits.

8.6.1 Adapting to Change

Agility and Flexibility: In a rapidly changing technological landscape, agility and flexibility are crucial for staying competitive.

- **Continuous Learning:**
    - **Stay Informed:** Regularly update your knowledge on the latest advancements in AI, quantum computing, and blockchain.

- **Adapt Skills:** Be prepared to learn new skills and pivot your career or business strategy as needed.

- **Flexible Work Practices:**

  - **Remote Work:** Embrace remote work and flexible schedules to attract and retain top talent.
  - **Cross-Functional Teams:** Create cross-functional teams that can quickly adapt to new challenges and opportunities.

Scenario Planning: Using scenario planning helps businesses and individuals anticipate future changes and prepare for different potential outcomes.

- **Identify Key Drivers:**

  - **Technological Trends:** Monitor key technological trends and their potential impact on your industry.
  - **Market Shifts:** Pay attention to market shifts and emerging customer needs.

- **Develop Scenarios:**

  - **Best-Case Scenario:** Plan for optimistic outcomes where technological advancements drive significant growth.
  - **Worst-Case Scenario:** Prepare for challenges such as regulatory hurdles, security threats, or market disruptions.
  - **Most Likely Scenario:** Create a balanced plan based on the most likely developments and trends.

8.6.2 Building Resilience

Risk Management: Developing robust risk management strategies is essential to navigate uncertainties and disruptions.

- **Identify Risks:**
  - **Technological Risks:** Assess the risks associated with adopting new technologies, such as cybersecurity threats and implementation challenges.
  - **Market Risks:** Evaluate potential market risks, including competition and changes in consumer behavior.

- **Mitigate Risks:**
  - **Contingency Plans:** Develop contingency plans to address identified risks and ensure business continuity.
  - **Insurance:** Consider insurance options to protect against significant financial losses related to technological disruptions.

Innovation Culture: Fostering a culture of innovation encourages experimentation and embraces change, driving long-term success.

- **Encourage Experimentation:**
  - **Pilot Projects:** Launch pilot projects to test new ideas and technologies before full-scale implementation.
  - **Failure Tolerance:** Promote a mindset that views failure as a learning opportunity rather than a setback.

- **Support Collaboration:**
  - **Open Innovation:** Encourage collaboration with external partners, including startups, academic institutions, and industry consortia.
  - **Interdisciplinary Teams:** Form interdisciplinary teams to leverage diverse perspectives and expertise.

8.6.3 Ethical Considerations

Responsible Use of Technology: Promoting the responsible and ethical use of AI, quantum computing, and blockchain is essential for sustainable development.

- **Ethical Guidelines:**
    - **Develop Standards:** Establish ethical guidelines for the development and deployment of new technologies.
    - **Transparency:** Ensure transparency in AI decision-making processes and blockchain transactions.
- **Ethical Training:**
    - **Employee Training:** Provide training programs on ethical considerations and responsible technology use.
    - **Stakeholder Engagement:** Engage stakeholders in discussions about the ethical implications of technological advancements.

Social Impact: Ensuring that technological advancements contribute positively to society and address pressing global challenges is a critical aspect of future-proofing.

- **Positive Contributions:**
    - **Sustainable Development Goals:** Align technological initiatives with the United Nations Sustainable Development Goals (SDGs) to address global challenges such as poverty, inequality, and climate change.
    - **Inclusive Growth:** Promote inclusive growth by ensuring that technological benefits are accessible to all segments of society.
- **Community Engagement:**

- **Public Involvement:** Involve the public in decision-making processes related to technology deployment, ensuring their voices are heard and considered.
- **Impact Assessments:** Conduct social impact assessments to evaluate the potential effects of new technologies on communities and address any negative consequences.

Conclusion: Future-proofing careers and businesses in the era of AI, quantum computing, and blockchain requires agility, resilience, and a commitment to ethical practices. By adapting to change, building robust risk management strategies, fostering a culture of innovation, and promoting the responsible use of technology, individuals and organizations can navigate uncertainties and leverage technological advancements for long-term success and positive societal impact.

This comprehensive guide provides practical strategies for future-proofing careers and businesses in the rapidly evolving technological landscape. By addressing various aspects of adaptation, resilience, and ethical considerations, readers can develop a holistic approach to navigating the challenges and opportunities presented by AI, quantum computing, and blockchain technologies.

8.7 Conclusion: Embracing the Future

The convergence of AI, quantum computing, and blockchain technology is set to redefine the way we live and work. As we stand on the brink of this technological revolution, it is imperative for individuals and businesses to prepare proactively. This concluding section recaps the key strategies discussed in this book and issues a call to action for embracing and driving innovation.

8.7.1 Summary of Key Points
Recap of Strategies:

- **For Businesses:**

    - **Strategic Planning and Roadmapping:** Develop comprehensive technology roadmaps that align with business goals and industry trends. Emphasize the importance of investment planning, identifying high-impact projects, and forming strategic alliances.

    - **Building a Tech-Savvy Workforce:** Conduct skills assessments, implement training programs, and promote continuous learning to upskill employees in AI, quantum computing, and blockchain.

    - **Fostering Innovation:** Encourage a culture of innovation by supporting experimentation, pilot projects, and collaboration with external partners. Invest in research and development to stay ahead of technological advancements.

    - **Risk Management and Resilience:** Develop robust risk management strategies, including contingency planning and adopting security best practices. Ensure compliance with regulations and standards.

- **For Individuals:**

- **Lifelong Learning:** Embrace continuous education by taking online courses and obtaining certifications in emerging technologies. Stay informed about the latest advancements through industry publications and research papers.
- **Networking and Community Engagement:** Join professional networks and participate in conferences, workshops, and webinars to learn from industry experts and connect with peers. Engage in online communities and forums.
- **Practical Experience:** Gain hands-on experience by working on real-world projects, contributing to open-source initiatives, and participating in hackathons and competitions.
- **Personal Development:** Develop soft skills such as critical thinking, problem-solving, and adaptability. Stay informed about technological trends and their implications for society.

Vision for the Future:

- **Transformative Potential:** The convergence of AI, quantum computing, and blockchain holds transformative potential for industries and society. These technologies will enable new business models, enhance productivity, and drive innovation across sectors.
- **Improving Lives:** By addressing global challenges such as healthcare, sustainability, and social equity, these technologies have the power to significantly improve the quality of life for people around the world. The

integration of AI, quantum computing, and blockchain will lead to smarter cities, personalized healthcare, and more efficient governance.

8.7.2 Call to Action

Encouraging Proactivity:

- **Take Proactive Steps:** Readers are encouraged to take proactive steps in learning about and adopting emerging technologies. This involves staying informed, seeking education and training opportunities, and applying new skills in practical settings.
- **Embrace Lifelong Learning:** Continuous learning is essential to staying ahead in the rapidly evolving technological landscape. Pursue courses, certifications, and hands-on experiences to deepen your understanding and expertise.

Embracing Innovation:

- **Drive Technological Advancements:** Individuals and businesses should embrace innovation and contribute to the technological advancements shaping the future. This includes participating in research and development, collaborating with peers, and sharing knowledge and best practices.
- **Foster a Culture of Innovation:** Cultivate a culture that encourages creativity, experimentation, and calculated risk-taking. Support initiatives that promote innovation within organizations and communities.

Empower Society:

- **Promote Ethical Technology Use:** Advocate for the responsible and ethical use of AI, quantum computing, and blockchain. Ensure that technological advancements are aligned with societal values and contribute positively to global challenges.
- **Engage with the Public:** Foster open dialogue between technologists, policymakers, and the public to build trust and understanding. Involve communities in decision-making processes to ensure that technological developments reflect the needs and values of society.

Conclusion:

The future is being shaped by the convergence of AI, quantum computing, and blockchain. By embracing these technologies, staying informed, and fostering innovation, individuals and businesses can drive transformative change and contribute to a better, more prosperous world. The journey ahead is filled with opportunities and challenges, but with proactive preparation and a commitment to ethical practices, we can harness the full potential of these advancements and create a future that benefits all of humanity.

www.ingramcontent.com/pod-product-compliance
Lightning Source LLC
Chambersburg PA
CBHW050047230526
45470CB00004B/1433